Brazil

Brazil

The Once and Future Country

Marshall C. Eakin

St. Martin's Griffin
New York

ISBN 0-312-21445-6 paperback
Library of Congress Cataloging-in-Publication Data
Eakin, Marshall C. (Marshall Craig), 1952-
    Brazil : the once and future country / Marshall C. Eakin
      p. cm.
    Includes bibliographical references and index.
    ISBN 0-312-16200-6 (cloth) 0-312-21445-6 (pbk)
    1. Brazil--Civilization--20th century.   I. Title.
F2537.E35 1996
981-dc20   96-28128
                                                     CIP

Design by Acme Art, Inc.

First published in hardcover in the United States of America in 1997
First St. Martin's paperback edition: October 1998
10 9 8 7 6 5 4 3

for
Michelle Beatty-Eakin
and
E. Bradford Burns
(1934-1995)

These people, the Brazilian people, intrigue
me with their generosity, their wisdom, and above
all, their unfailing hope.

–Roberto DaMatta

CONTENTS

Source: Werner Baer, *The Brazilian Economy: Growth and Development*, *4th ed.* (Westport, CT: Praeger, 1995).

PREFACE

"Brazil," Antônio Carlos Jobim once remarked, "is not for beginners." I wrote this book in the belief that Brazil's great composer was wrong (at least on this point), and that we all embark upon the path to cross-cultural understanding as beginners. Brazil is one of the great countries in the world today. It is also one of the great unknowns to most of the English-speaking world. I hope this book will, in some small way, begin to change this by introducing those who know little about Brazil to its beauty and wonders as well as its problems and shortcomings. This is a book for beginners.

The idea for this book has taken shape over two decades as I have tried to formulate my own understanding of Brazil and its paradoxes and contrasts. While I have written and taught about Brazil over the past twenty years, countless friends and acquaintances have asked me for *one* good book they could read to prepare them for their first visit to Brazil. Unable to offer an adequate response, I decided to write that book. Although the English-language literature on Brazil is vast (especially travel accounts about the Amazon), the best (and nearly the only) introduction to Brazil in English dates from the early 1970s, when the anthropologist Charles Wagley published a revised edition of his *Introduction to Brazil* (originally published in 1963). Building on decades of writing by Brazilians, Americans, and Europeans, this book offers a concise and comprehensive introduction to Brazil's history, regions, society, culture, politics, and economy.

I am acutely aware that conveying the complexity and diversity of Brazil to non-Brazilians is not an easy task. In essence, I am a cultural broker, much like an anthropologist, acting as an intermediary between two cultures. This book represents my efforts to translate the world and worldview of Brazilians into terms comprehensible to others. Inevitably, through the process of translation and generalization I simplify and, no doubt, distort some aspects of Brazil and its people. I understand and acknowledge that perfect translation does not exist in literature or cross-cultural communication. Like the anthropologist, I hope that I have come to understand Brazil as well as an outsider can, while I strive to communicate that understanding as best I can to an English-speaking audience.

My own effort to understand Latin America began some twenty-seven years ago when I was a high school student in Houston, Texas. A summer doing public health work among the Maya Indians of the Guatemalan highlands in 1970 sparked a love affair with Latin America that has intensified as the years have gone by. A year-and-a-half (1973-74) as an exchange student at the University of Costa Rica, and the decision to do doctoral work in Latin American history, turned the affair into something of a marriage. While I was in graduate school at the University of Kansas and UCLA in the late seventies, my principal interest gradually shifted from Central America to Brazil. Over the last eighteen years, I have made nine extended trips to live and work in Brazil that have provided the raw material for my writing and teaching. My position as a professor of history over the past decade has given me the luxury and good fortune to make a career pursuing my passion for Latin America.

As a historian, I am committed to the notion that the workings of the modern world cannot be understood unless they are placed in historical perspective. My occupational bias permeates this book. History receives top billing. Chapter 1 provides the reader with the historical context behind the analysis in the following chapters. The historical survey will the give the reader a sense of the gradual movement of peoples into the diverse regions that make up Brazil. Like other nations of continental dimensions, Brazil is a country of markedly distinct regions, and chapter 2 presents an introduction to the "nations within a nation." Chapter 3, in which I attempt to explain the principal characteristics of Brazilian society and culture, is perhaps the most audacious section of the book. Chapter 4 dissects the Brazilian political system, and chapter 5 examines the economy.

In the conclusion, I return to the central theme of this book—Brazil's efforts to escape the pessimistic analysis of the popular proverb about the country of the future that appears at the beginning of the introduction. Brazilians have created a rich and complex culture, an industrial economy, and one of the world's largest democracies. Despite these impressive accomplishments, Brazil has yet to realize its enormous potential. In the following chapters, I discuss Brazil's achievements and its failures, its merits and flaws, in the hopes that I can bring Brazil to life for non-Brazilians. Although I continually compare and contrast Brazil with the United States and Western Europe, my intent is not to show that somehow Brazil fails to measure up to these nations or to say that it should be like them. Brazil fascinates and attracts the foreigner precisely because it is not the United States or Europe. As descendants of Europeans, Africans, and Native Americans, Brazilians have their own unique past and their own path to the future. I hope that this book will convey the essence of that past, of the country Brazil once was, the nation it has become, and the promise of the country it could be. I hope it will show the reader why I, along with Roberto DaMatta, find the Brazilians so appealing and intriguing.

ACKNOWLEDGEMENTS

One of the great pleasures of finishing a book is the opportunity to thank the many people who helped make it possible. My greatest debt is to the many Brazilians over the past twenty years who have put up with my incessant questions about every facet of their country and culture. I am especially indebted to the people of Minas Gerais, and more specifically, Belo Horizonte, my Brazilian home away from home since 1979. Brazilians and Brazilianists alike will, no doubt, see my interpretation of Brazil as skewed by the perspective of a *mineiro-americano*. My special thanks go to Yonne Grossi, Amilcar and Roberto Martins, Virgílio and Rejane Almeida, and Douglas Libby in Belo Horizonte; and Vera and José Ribamar Gomes in Rio de Janeiro. *Obrigado por tudo.*

Malcolm Silverman enticed me into studying Portuguese more than twenty years ago while I was a very confused undergraduate exchange student trying to learn Spanish at the Universidad de Costa Rica. Little did he (or I) suspect that he had initiated my career as a Brazilianist, pointing me in the right direction (further south and east). Charley Stansifer and Bill Griffith turned me into a historian of Latin America, and Betsy Kuznesof came along at just the right time to convince me that Brazil really was the country of *my* future. I will be forever grateful for their guidance.

Over the years, I have been fortunate enough to receive a series of grants and fellowships that have allowed me to return to Brazil repeatedly. Although these agencies thought they were funding highly specialized research, they got much more than academic publications for their investment. The National Defense Education Act (Title VI), the University of Kansas, UCLA, the Fulbright-Hays program, the Tinker Foundation, the Fundação João Pinheiro, Vanderbilt University, and the National Endowment for the Humanities helped make this book possible.

For the past fourteen years, I have been lucky to work with an exceptional group of colleagues in the Department of History and the Center for Latin American and Iberian Studies at Vanderbilt University.

In particular, Simon Collier and Jane Landers have provided me with friendship, support, and encouragement. A strong and supportive group of Brazilians and Brazilianists have surrounded me with a little of Brazil in Nashville—the late Alex Severino, Rodolfo and Rosa Franconi, Margo Milleret, Russell Hamilton, Jim Lang and Cecilia Grespan, Keo and Aline Cavalcanti, Sam Morley, Wendy Hunter, Kurt Weyland, Talisman Ford, Beth Conklin, and Tom Gregor. *Um muito obrigado especial* to Norma Antillón and Paula Covington. Frank Wcislo, as a good Russianist, helped keep me going with his comradeship. A special thanks goes to Mona Frederick, of the Robert Penn Warren Center for the Humanities, who provided me with a wonderful home while I wrote the first draft of this book.

A number of colleagues read and commented on earlier drafts of this work. My thanks go out to Simon Collier, Todd Diacon, Wendy Hunter, Kurt Weyland, and Joel Wolfe for providing me honest critiques that often made me rethink and sharpen my analysis. My students at Vanderbilt have shaped this book in ways they never imagined with their questions and comments. They have long been the testing ground for the ideas and structure of this work.

Michelle Beatty-Eakin and E. Bradford Burns have done more to make this book possible than anyone else. My partner and best friend, Michelle has shared my obsession with Brazil for more than twenty years. As we have helped each other through graduate school; career changes; living in Costa Rica, Kansas, California, Tennessee, and Brazil; and rearing our two strong-willed daughters, she has been unfailingly supportive, encouraging, and understanding of my need to travel and write about Brazil. (And, yes, Lee and Lacy, dad's book is finally finished!)

My greatest regret about this book is that I did not finish it soon enough for Brad Burns to see it. A historian with a broad and sweeping vision of Latin America, Brad never shied from looking at the "big picture" or from trying to reach audiences beyond the academy. Beginning in the 1960s, he played a prominent and pioneering role in promoting the study of Brazil in the United States. Brad was also a wonderful and talented writer, teacher, and mentor. I dedicate this book to Brad and Michelle, the two most important influences on my own education and the writing of this book.

—Belo Horizonte, Minas Gerais
Ash Wednesday, 12 February 1997

The Country of the Future?

> Brazil is the country of the future—and always
> will be.
> —popular Brazilian proverb

WITH HALF OF SOUTH AMERICA'S LAND and one-third of the population of all Latin Americans, Brazil is a land of the marvelous and the
mystical, the sublime and the tragic. Contrasts and paradoxes define
Brazil: immense wealth surrounded by widespread poverty, a modern
industrial infrastructure alongside an outmoded agricultural system, a
dynamic and largely white South and a northeastern coast that is
overwhelmingly of African descent. Brazil combines the features of both
the most advanced and sophisticated industrial nations of the Northern
Hemisphere as well as characteristics of the poor agricultural societies of
the so-called Third World. In the 1970s, one Brazilian economist coyly
suggested that the country be renamed Belindia—because its modern
industrial base resembled Belgium, while its backward social structure
looked more like India.

A nation of huge dimensions, Brazil is the fifth largest country in
the world. At just over 3.3 million square miles, it is slightly larger than
the continental United States. Nearly 2,700 miles across at its widest
extension from the western Amazon to the Atlantic, the country spans
nearly 2,500 miles from the dense jungle on its northern boundary with
Venezuela to the expansive prairies of Uruguay and Argentina in the
south. Its 155 million inhabitants make Brazil the fifth most populous
country on earth, and the largest Catholic nation in the world. Brazil
now has the tenth largest economy in the world (with a Gross Domestic
Product of $688 billion in 1995), placing it just behind the seven major

industrial powers (the United States, Japan, Germany, France, Italy, the United Kingdom, Canada), Russia, and Spain. Within the next decade, it could surpass Russia and Spain. With more than 90 million voters participating in its 1994 presidential elections, Brazil is the third largest democracy in the world (after India and the United States).

Although Brazil shares borders with all but two of the other eleven independent nations of South America, most of its border areas are far from the major population centers of the eastern coast. For centuries, the country's major cities have been concentrated in the Southeast and Northeast and oriented toward the Atlantic. Brazil, its back to the rest of South America, faces Europe and Africa. In fact, the major cities of northeastern Brazil are physically closer to West Africa than to neighboring Peru and Colombia.

History, geography, culture, language, and economics have reinforced the rise of regional identities within the Brazilian nation. One of the continuing themes in Brazilian history has been the struggle to forge these regions into a single nation, a single people. In the sixteenth century, the Portuguese carved out a colonial plantation society in the Northeast centered around the present states of Bahia and Pernambuco. Plantation agriculture and slavery left a deep imprint on the coastal zone which today is the most Africanized region of the country. In the interior of the Northeast, the racially mixed descendants of Indians and Europeans have struggled for centuries to eke out an existence in the arid backlands. With the discovery of gold in the mountains of the Southeast at the end of the seventeenth century, the Portuguese for the first time moved inland and effectively occupied a piece of the interior. As the Southeast and its principal port of Rio de Janeiro boomed in the eighteenth century, the Northeast and the colonial capital of Salvador entered into a long economic decline from which it has only recently begun to emerge. Since the eighteenth century, the Southeast—and its key states of Rio de Janeiro, São Paulo and Minas Gerais—has formed the economic, political, and population core of the country. In this powerful heartland, nearly 43 percent of the population occupies just over 10 percent of the country's landmass and generates 65 percent of the nation's wealth.

The South, stretching from São Paulo to the borders with Argentina and Uruguay, took shape with a wave of European immigration in the nineteenth century. Today the South is ethnically the most European section of the country, and economically it is second only to the Southeast. Cattle ranching, modern agriculture, and industry have forged a region with some of the highest living standards in Brazil. The settlement and exploitation of the North and the Center-West have been taking place in the last generation. Although it takes up about 40 percent of the nation's landmass, the North, home to the vast Amazon River Valley, remains sparsely populated. The inauguration of the new national capital at Brasília in 1960 initiated a movement into the Center-West that accelerated in the 1970s and 1980s. Still very much a frontier society, the Center-West stretches from the southern Amazon to the borders with Paraguay and Bolivia. Together, the North and Center-West comprise nearly two-thirds of Brazil's land mass and just 13 percent of its population.

Recent politics highlight both the promise and the problems facing the Brazilian people. On October 3, 1994, more than 90 million Brazilians cast their ballots in national elections. Fernando Henrique Cardoso, a sociologist, former political exile, and recent finance minister, won the presidency with 54 percent of the vote. During mid-1994, Cardoso managed to forge a coalition of political parties that provided him with the largest winning percentage in modern Brazilian politics. A social democrat, Cardoso assumed the presidency on January 1, 1995. His four-year term began amidst enormous problems, as well as an atmosphere of high expectations, hope, and optimism following fifteen very difficult years for most Brazilians.

After a decade of economic crises characterized by enormous foreign debt ($120 billion), runaway inflation (over 50 percent per *month* at times), increasing social unrest, and seemingly endless political scandals, Cardoso faces major challenges. His most important challenge is to persuade the diverse regional and political interests to restore their faith in the political process and democratic institutions. Political polarization, economic crisis, and social unrest led to the collapse of democratic politics in 1964, when the military staged a coup and took power. The authoritarian generals did not return power to civilians until 1985, and then only after an indirect presidential election (in the carefully selected electoral college). In 1989, more than 80 million Brazilians voted in the first direct presidential election since 1960. They elected Fernando Collor de Mello. His impeachment in 1991, and subsequent investigations of graft and influence peddling in congress and the government, have produced widespread cynicism in Brazil about the entire political process.

Much of the cynicism stems from the severe economic problems the country has experienced during the past fifteen years, after more than a century of impressive economic growth. A nation of enormous resources and a huge economy, Brazil has long seemed to many to be destined to be a power in the modern world. Yet, during the 1980s the country experienced recession after recession, a severe debt crisis, and runaway inflation. Many Brazilians began to wonder openly if Brazil was condemned to remain a country of unfulfilled potential.

By the 1980s, Brazil had become one of the world's leading manufacturers of automobiles, armaments, and aircraft, with an industrial base closer to that of Europe than most of Latin America. With a population of some 15 million, the cosmopolitan city of São Paulo has become the largest industrial center in the developing world and the second largest metropolis in the Americas (after Mexico City). The Brazilian government has been developing its own missile and space program, as well as a controversial nuclear plan. Brazil has the world's eighth largest market for computers, and the government has promoted the growth of a national computer industry to challenge imports from the United States, Europe, and Japan.

With the world's largest reserves of iron ore, and impressive deposits of uranium, bauxite, manganese, and precious gems, Brazil has exceptional natural resource potential. (Brazil, however, has very little coal and unimpressive oil deposits.) The world's leading producer of coffee since the nineteenth century, and the leading exporter of sugar, Brazil also has become the world's fourth leading exporter of agricultural products. Russians and Americans consume Brazilian oranges, Iraqis eat Brazilian chickens, and the Japanese import huge quantities of Brazilian soybeans.

The oil crisis of the seventies, rising interest rates, and an enormous foreign debt brought the economic surge of the 1960s and 1970s to a grinding halt. A new economic plan that has stabilized the currency (the *real*), the election of the plan's architect (President Cardoso), and dramatically reduced inflation rates (from 50 to 1 percent per month) have revived the Brazilian economy and the people's hopes. Once again, as in the 1950s and 1970s, the Brazilian economy appears on the verge of another phase of impressive economic expansion. Once again, optimism has begun to return to Brazil.

Yet, beneath the modernity of industrial and agricultural expansion, Brazil retains many of the structures and forms of a traditional society. The modern industry and skyscrapers of São Paulo, Rio de Janeiro, and Brasília contrast sharply with the poverty of the backward and economically depressed rural life in the Northeast. The shanty towns (*favelas*) of Rio and São Paulo alone contain some 4 million Brazilians. Despite the economic growth of the past half-century, two-thirds of the population live below the officially defined poverty level of roughly $200 a month. The gap between rich and poor intensified between 1960 and 1990, even as the economy grew faster than any other in the world. According to a 1995 World Bank study, Brazil has the most unequal distribution of wealth of any country in the world. Brazil remains a rich country full of poor people.

Further compounding the contrasts and paradoxes is the powerful and pervasive mixture of Western and non-Western culture. A devoutly Catholic Portugal conquered and forcibly converted the Native American population; it then, for four centuries, imported millions of African slaves to labor on plantations and in mines. The three races intermixed and produced a racial and cultural melting pot that startles most foreigners, especially North Americans who are accustomed to a society with sharp divisions between blacks and whites. The impact of Africa on Brazilian society and culture astounds and fascinates foreigners. The collision of Native American, European, and African races and cultures produced a unique society with some salient non-Western cultural traditions. Most notably, African worship of spirits blended with Catholic veneration of saints to produce a form of spiritism that coexists with Roman Catholicism, and that is widely accepted throughout Brazilian society. An old saying sums up this collision of three worlds: "Brazil's land is American, its facade Iberian, and its soul African."

An emerging industrial power of continental dimensions, with strong traces of its colonial plantation heritage, and blending Western and non-Western culture, Brazil presents a remarkable and important challenge that outsiders need to understand. Brazil stands poised between two worlds and two eras. While it does not yet belong among the leading nations of the industrial world, it has emerged as a middle power, and it has the potential to become a world power in the next century. A nation born out of the European imperial expansion after 1500, Brazil continues to bear the burdens of its colonial heritage as it pursues modernity. Caught between past and present, tradition and modernity, Brazil provides us with a glimpse of the slow death of one era of world history and the even slower birth of a new world order whose outlines are still blurred and confusing. We must turn to Brazil's rich history—to the country Brazil once was—to understand how this complex and paradoxical blend took shape, and how it shapes Brazil's future.

ONE

The Presence of the Past

I live in the twentieth century,
I'm off to the twenty-first,
Still the prisoner of the nineteenth.
      —Affonso Romano de Sant'Anna

ALL NATIONS CARRY WITH THEM the scars of their past, yet Brazil bears the "burdens of history" more visibly than most. Everywhere one looks in Brazil, the past intrudes upon the present. The modern, rapidly changing "country of the future" appears unable to escape another traditional and unchanging Brazil that is seemingly frozen in time. This contrast has misled many to argue that two separate societies coexist in Brazil. In fact, what seem to be dual societies are not. Both traditional and modern, past and present, interpenetrate. As the anthropologist Roberto DaMatta has observed, Brazilians "have not abandoned the past and yet embrace the future with all their might." "Tradition" and "modernity" represent different facets of the same society. The traditional side of Brazilian society is intimately bound up with the modern side. The "modern" features may tell us what the future holds for Brazil. The "traditional" features of society reflect the historical path Brazil has taken over the past five centuries. To understand the bonds between traditional and modern Brazil, we must retrace that path.

A violent and prolonged cultural clash between peoples from Europe, Africa, and the Americas shaped the formation of Brazilian society. Peoples from what came to be called the "Old World" crossed the Atlantic conquering the indigenous cultures in lands the Europeans labeled the "New World." When these peoples could not, or would not, provide the labor the conquerors demanded, the

Europeans forcibly removed millions of Africans from their homelands and reduced them to slavery in the Americas. This complex
and brutal clash of Europeans, Native Americans, and Africans
followed similar patterns in the Caribbean, as well as the areas that
were to become the southern United States and eastern Brazil. The
Europeans imposed their political order, languages, and economic
system on the peoples of the Americas. The vanquished resisted
with ferocity and determination, both physical and cultural,
transforming the institutions of the Europeans and ultimately
creating societies that were neither European, nor Native American, nor African.

Europeans settled in Brazil and constructed a colonial society
that was in constant struggle with Amerindians and Africans. The
colonists eventually broke with their mother country and established political independence. Over the past two hundred years,
Brazilians have forged a nation-state out of an immense and
imposing landscape, and a rainbow of peoples and cultures. They
have created a racially mixed society with enormous economic
inequities; a highly centralized state with marked regional identities; and a national culture that binds together peoples from an
array of ethnic, linguistic, and cultural traditions. This process was
neither predetermined nor inevitable. The history that unfolds in
the following pages traces the creation and construction of Brazil
from a small series of colonial outposts into a modern nation.

THE RISE OF THE WEST

The history of the Americas makes no sense unless it is placed within the
larger context of European overseas expansion. Beginning in the fifteenth century, the Europeans crisscrossed the globe, eventually subdu-
ing the peoples they encountered and establishing the global supremacy
of the West. The nations of the Western Hemisphere began as colonial
enterprises, and Europe played the central role in the formation and
development of New World societies for centuries. Today, the Americas
continue to grapple with the legacies of colonialism. European expansion eventually created a truly global economic system that linked up all
societies and cultures. Brazilian history began in the sixteenth century,

when a group of small, newly emerging nations in Europe embarked on overseas voyages in pursuit of trade with Africa and Asia.

The rise of the West to world domination from the fifteenth to the nineteenth centuries was by no means an inevitable process. Prior to 1500, the peoples of the globe were scattered and relatively isolated, and Europe was but one of many power centers in the world. The Chinese had constructed a highly developed civilization over millennia in East Asia, and the Moguls ruled over a vast empire on the Indian subcontinent. The Ottoman dynasty controlled much of North Africa, the Middle East, and southeastern Europe, while the Persian empire stretched from present-day Iran across the Arabian peninsula. Russian rulers had forged a powerful kingdom that was centered in Moscow. The great trade routes connected Europe with Asia via Moslem merchants in the Middle East, and black Africa with Europe via Moslem traders in North Africa. Europeans had only vague and fanciful ideas of the dimensions of the lands beyond the Mediterranean rim, mostly from the writings of a few adventurous travelers like Marco Polo. On the other side of the Atlantic, the Aztecs dominated central Mexico and the Incas controlled an empire in the Andes. While all these empires had expanded and gained control of vast territories, only the Europeans with their newly emerging and fragile monarchies moved outward across the seas.

Complex and interacting factors drove the Europeans outward and made their success possible. By the late fifteenth century, the first of the new nation-states had begun to consolidate around a handful of the numerous and dispersed feudal monarchies. Portugal (in the twelfth century) and Spain (in the fifteenth century) pioneered this process of state-building. England, France, and the Netherlands followed in the sixteenth century. Once these monarchies had established centralized authority and consolidated control over the land, they could then turn their attention to overseas expansion. The continual competition among these relatively small states created a dynamic of innovation in politics, society, and economics.

Innovations in science and technology provided the new nations with the means for expansion. Europeans gradually assimilated astronomical knowledge, as well as the compass and navigational tools, that had been developed in China, India, and the Middle East. By the late fifteenth century, they had also developed the ships that would take them around the globe—most notably, the caravel. Unlike the ancient, square-masted Roman galleys that plied the relatively calm Mediterranean, the caravel had a potent combination of technology: a large hull for sailing the high seas, a sternpost rudder and triangular sails for directional mobility, and artillery to intimidate those who challenged or refused to cooperate with them. Knowledge of the stars and the creation of navigational instruments gave the Europeans the means to chart their course. The caravel provided them with the means to sail it.

Science, technology, and the consolidation of nation-states paved the path to expansion; cultural and economic forces propelled Europeans down the path. Western culture is peculiar in its emphasis on the systematic pursuit of practical knowledge about the natural world. Many civilizations through the ages have contributed to the growth of science and technology, but only the West brought together the disparate contributions of other civilizations to produce the revolution that gave birth to modern science in the sixteenth century. Only in the West was technological innovation so aggressively promoted and revolutionized. The pursuit of scientific and technological advances became both an end in itself and the means to other ends. Science and technology became instruments for expansion and colonial exploitation.

Christianity played an even more important role in Western culture and its expansion. With the prominent exception of the Jews (who were concentrated mainly in Eastern Europe), Europe was a Christian civilization. And in 1500 there was but one church, one christianity, in Western Europe—the Catholic ("universal") Church headquartered in Rome. Although bloody wars dramatically and irrevocably shattered this religious unity in the sixteenth century, Protestants and (and even more so) the Catholics shared an aggressive zeal to convert all peoples and to save their souls. In large part due to a centuries-long battle to drive the Moslems from Spain and Portugal, Iberian Catholicism became the most militant and aggressive Catholicism in Europe. Although often mixed with other motives, the desire of Europeans to find and save new souls was sincere and profound. Overseas expansion offered an unprecedented opportunity to carry the message of Christ to other peoples. Bernal Díaz del Castillo, one of the conquistadors of Mexico, captured this mixture of motives when he observed that he and his compatriots embarked upon their mission "in the service of God and

His Majesty, and to give light to those who sat in darkness—and also to acquire that gold which most men covet." The spiritual conquest played a central role in the construction of Spanish American and Brazilian culture. The Iberians set out to conquer the Native Americans with both the sword and the cross.

Greed was also a powerful motive driving Western expansion, and the emergence of capitalism pushed the Europeans outward. By 1500, the complex and growing trading system in Europe had laid the basis for the birth of a capitalist economy driven by the pursuit of profit. And by 1500, the most impressive profits came from the trade of luxury goods and rare spices from the East. Middle Eastern and North African intermediaries stood in the way of the Europeans and the suppliers. Given the difficulties (vividly demonstrated by the Crusades in the late Middle Ages) of trying to wrest control of the overland trade from the intermediaries by force, the logical alternative was to establish sea routes to the East. Portugal, Europe's "window to the Atlantic," was strategically positioned to serve as an intermediary in the trade between the Mediterranean and northern Europe, and to become a jumping off point to the East.

Pursuit of profit and souls drove the Europeans out onto the high seas and around the globe while the new science, technology, and monarchies provided them with the means to find their way. In the fourteenth and fifteenth centuries, the small city-states of northern Italy pioneered the expansion. Strategically located between Europe (which was experiencing economic growth) and the Eastern overland routes, Venice, Florence, and Genoa had developed sophisticated trading economies and pioneered credit and financial tools such as double-entry bookkeeping and banking. In the sixteenth century, Portugal and Spain, the first of the centralized monarchies, left the Italian city-states behind. After 1492, the Iberian empires in the Americas and the East shifted the axis of the European economy out of the Mediterranean and into the Atlantic for the first time in history.

TWO WORLDS COLLIDE

By 1492, Africans, Asians, and Europeans had some knowledge—however vague and fanciful—of each other. The peoples of the New

World and the Old, however, were unaware of each other's existence. Ironically, in their pursuit of a route to Asia, the Europeans stumbled upon the Americas, bringing ten thousand years of mutual isolation to an end. A sixteenth-century Spanish chronicler who recognized the profound impact of the joining of the two worlds called the European discovery of the Americas "the greatest event since the creation of the world (excluding the incarnation of Him who created it)." The voyage of Columbus initiated an irreversible exchange of plants, animals, diseases, and peoples that transformed the planet. The "Columbian exchange" also unleashed a cultural and biological maelstrom that would enrich the Europeans and devastate Native Americans.

The Americas had been populated as a result of waves of Asian migrations across what are now the Bering Straits that began about 20,000 years ago and continued up to 4,000 years ago, when the Eskimos (Inuit) occupied areas in the Arctic Circle. Over millennia, cultures of widely divergent levels had coalesced. They ranged from nomadic hunting and gathering tribes to the Aztec, Maya, and Inca empires whose civilizations rivaled those of the Old World. Although these Asiatic migrants eventually developed many languages, cultures, and civilizations, biologically they formed the most homogeneous large population on the planet. In particular, they formed an enormous gene pool that had never been exposed to the diseases of the Old World.

Smallpox, measles, and influenza provoked a demographic catastrophe of unprecedented proportions in human history. Within the first fifty years after the initial voyage of Columbus, the native population of the Americas declined by possibly 80-90 percent. Epidemics swept through Native American societies, killing entire villages, clearing the path for the European conquest, and provoking cataclysmic cultural disorientation and disruption. The military conquest wreaked havoc among Native American peoples, but the biological conquest nearly annihilated them.

In the Americas, the nature of the European conquest in each region—and the evolution of each colonial society—to a large degree depended on the kind of society the Europeans confronted. The Spanish conquistadors encountered densely populated and complex empires in central Mexico and Peru. After defeating the Incas and Aztecs, the Spanish monarchy replaced the deposed native elites with its own royal representatives. Spain superimposed its own imperial system on that of the Incas and Aztecs, as the masses continued to labor and pay tribute, but to new rulers.

The Tupí-Guaraní cultures of eastern South America did not construct imposing empires. The hundreds of distinct tribes generally subsisted off hunting and gathering or semi-sedentary agriculture. Even in the latter case, the primitive slash-and-burn agriculture did not produce enough wealth for the construction of cities or complex economic and political systems. These tribes had no metal tools, no written language, and no beasts of burden, and they (like people in the rest of the Americas) had not discovered the use of the wheel. They worshipped spirits and relied on shamans (spiritual leaders) to help influence the spirits and to interpret the meaning of natural and supernatural signs. Unlike the densely populated empires of Mexico and Peru (which possibly approached a combined population of 50 million inhabitants), the native population of eastern South America is estimated to have been approximately 3-6 million in 1500.

The Spanish and the Portuguese arrived in the Americas by mistake on their way to India and China. Portuguese explorers had been leading the way in overseas expeditions for most of the fifteenth century. Spurred on by the legendary Henry the Navigator in the early 1400s, Portuguese expeditions had charted the west coast of Africa as far south as present-day Angola by the 1480s. In 1488, an expedition led by Bartolomeu Dias rounded the Cape of Good Hope (as the Portuguese later christened it) and entered the Indian Ocean before a mutiny by his crew forced Dias to turn back. Dias proved for the first time that Europeans could travel around Africa by sea and establish direct trade with the East. Like Moses, Dias saw the promised land but would not be allowed to enter.

As legend has it, standing on the docks in Lisbon witnessing the return of Dias' ships was a young Italian from Genoa who would also alter the course of world history. Christopher Columbus had supposedly survived a shipwreck off the coast of Portugal, stayed on to learn all he could of navigation and sailing, and gained experience on Portuguese expeditions in West Africa. Possessed by an almost mystical belief, and armed with erroneous calculations, he tried to persuade the Portuguese monarchy to back his plan to reach India by sailing due west. All educated Europeans believed the earth was round, and the Portuguese knew that Columbus had severely underestimated the circumference of the globe. They also believed that the eastward route to India and China presented fewer problems than a westward voyage into uncharted waters. Eventually, Columbus gained the backing of Isabel, the queen of Spain. In 1492, he reached what by his calculations should have been China and Japan. He reported back about his discovery of the "Indies" and the "Indians" who lived there (creating problems of labelling that still plague us).

Columbus stopped off in Portugal on his way back to Spain and reported his discovery to the incredulous Portuguese, who reasonably enough thought him to be a bit mad. They knew, after all, that Columbus could not have reached the Indies given the distances he sailed, and besides, Dias had already shown the correct way to India. Spain immediately sought papal and international recognition of its "rights" to the newly discovered lands. In 1494, with the pope's blessing, Spain and Portugal signed the Treaty of Tordesillas, which drew an imaginary line far out into the Atlantic. With a few exceptions, the Portuguese would lay claim to colonial territories to the east, and Spain to the west, of the line. (Francis I of France acidly commented that he would like to see the clause in Adam's will where God divided the world between the Spanish and the Portuguese.)

The Portuguese kept on their eastward mission, and in 1498 an expedition led by Vasco da Gama proved them right. He reached India and returned with goods that produced a 700 percent profit for the expedition's financial backers. Da Gama opened Asia to European penetration and to the highly lucrative spice trade. Meanwhile, Columbus made three more voyages to the Caribbean, convinced that he had reached the Far East and frustrated by his inability to prove it.

As soon as Da Gama returned to Lisbon in 1499, the crown began preparing a second expedition. More than a dozen ships commanded by Pedro Alvares Cabral, a thirty-year-old nobleman, left Lisbon in March 1500. Following the procedure of his predecessors, Cabral guided his fleet in a southwesterly direction far out into the Atlantic, to catch the prevailing winds that would carry his fleet eastward to the Cape. For still unknown reasons, the fleet veered too far west and sighted land on April 22. Thus began the Portuguese presence in the Americas. Rarely has a geographical discovery that seemed so insignificant at first ultimately proved so important. Cabral had stumbled upon a continent—and Portugal's future.

The seeming innocence and simplicity of the local peoples awed and amazed Cabral and his men. These people appeared to him to be noble savages in some kind of Garden of Eden. The trip's chronicler reported back to the king that the natives were "naked and without any covering: they pay no more attention to concealing or exposing their private parts than to showing their faces, and in this respect they are very innocent." After nine days reconnoitering the coast and observing the local peoples, Cabral set sail for India and sent one ship back to Portugal to report his discovery to the king. Although optimistic about the possibility of converting such a "simple" people to Christianity, Cabral had seen little to stimulate an economic interest in what he christened the "Land of the True Cross."

Spices and luxury goods from the East had attracted the Portuguese for decades, and Cabral would continue the process Da Gama began of pursuing trade in the Indian Ocean. This prompted Francis I of France to refer contemptuously to his contemporary, Manoel I of Portugal (1495-1521), as the "grocer king." In just two decades, the Portuguese took control of this lucrative trade and established the world's first truly global empire. They constructed a commercial and maritime network that spanned four continents. While the Portuguese built their empire in the East, the Spanish focused their attention on the Western Hemisphere. Through extraordinary luck, technological superiority, and sheer audacity, the Spanish subdued not only the peoples of the Caribbean and northern South America, but also the empires of the Aztecs and the Incas. Fortunately for Spain, both central Mexico and the Andes contained not only dense populations but also rich silver mines that would make Spanish kings the most powerful in the world for nearly a century. With no apparent mineral riches and no dense population of laborers to exploit, the Portuguese understandably looked first to their empire in the East and paid little attention to Cabral's "island" in the Atlantic.

FROM BARTER TO SLAVERY

Following their pattern in the East, the Portuguese established trading posts (or factories) along more than 1,000 miles of the South American coastline. (Fortunately for the Portuguese, most of eastern South

America, as it turned out, lay to the east of the Tordesillas demarcation line.) Portuguese traders visited the factories with some frequency, mainly to load cargoes of a hardwood that produced a red dye known by its Latin name of *brasile*. Eventually, the land became identified with the brazilwood it produced, and the Portuguese began to call their small colony Brazil.

French attacks and incursions along the coast threatened Portugal's claims in South America, and the monarchy decided to send out an expedition in 1530 to establish a more permanent presence. Led by Martim Afonso de Sousa, the expedition marked a turning point in Portugal's role in Brazil and a departure from its imperial practices. Sousa founded the settlement of São Vicente (near present-day Santos), introducing sugar cultivation, cattle raising, and a colonial bureaucratic presence. The crown attempted to divide up 2,500 miles of coastline into a dozen captaincies, granting control of these new territories to members of the nobility. In exchange for developing and protecting their captaincies, these nobles (known as *donatários*) received control over domains that were sometimes larger than Portugal itself. The colonization scheme departed from the imperial practice of creating a network of maritime trading centers. Only in the small islands in the Atlantic (Madeira, the Azores, Cape Verde) had the Portuguese tried colonization. These tiny islands would eventually serve as the prototype for a Brazilian colonial experiment on a vast scale.

Many of the *donatários* never even saw their land grants. Four were not even settled, and only two (São Vicente in the South and Pernambuco in the North) experienced any initial success. French attacks continued to challenge Portugal's tenuous control of the coast, and the colonists faced a hostile or uncooperative indigenous population. When the settlers tried to enslave them, the Indians fought back, sometimes destroying the fragile settlements. In some captaincies, the colonists established trading relations with the Indians through barter arrangements, but this system never provided a stable and dependable work force for the small but growing agricultural settlements.

By mid-century, the crown had decided to impose a stronger royal presence and a more permanent agricultural base in the colony. In 1549, an expedition of some 1,200 colonists, soldiers, priests, and royal officials led by Tomé de Sousa established a permanent colonial capital on a strategically sound and beautiful location overlooking the Bay of All

Saints of the Savior (*Bahia de Todos os Santos do Salvador*). (Since it is known as both Bahia and Salvador, I will use the latter when referring to the city and the former when discussing the surrounding captaincy.) Within two decades, the sugar cane that the colonists had transplanted from Portuguese islands off the West African coast had spread rapidly in the rich soils of the hinterland around Salvador.

Long cultivated in the Mediterranean, sugar cane had been transplanted by the Portuguese to the newly conquered Azores and Madeira in the 1400s. Using slaves from the African mainland, they had established plantations and exported sugar, making it readily available for the first time in Europe. The crown encouraged the expansion of sugar cultivation in Brazil as a way of turning the poor colony into a revenue-producing venture. As cultivation spread, the demand for labor increased, and conflict between Indians and colonists intensified.

When barter proved incapable of generating a dependable labor force, the Portuguese turned increasingly to slavery as an alternative. In both the North (Pernambuco and Bahia) and the South (São Vicente) colonists fought, captured, and enslaved Indians to satisfy the rising demand for labor. As Father Antônio Vieira later lamented, the colonists captured the Indians "to draw from their veins the red gold which has always been the mine of that province!" Old World diseases (smallpox, measles, influenza), previously unknown in the Americas, decimated the population. Ironically, the demographic catastrophe intensified the pressure on survivors as the size of the potential labor pool diminished.

The survivors resisted Portuguese demands for labor by retreating into the interior, or by attacking the colonists. They also found a staunch ally in the Catholic Church, or more precisely, in the Society of Jesus. Six Jesuits had disembarked at Salvador in 1549, and they effectively founded the institutional presence of the Catholic Church in Brazil. The Jesuits were a very new and aggressive religious order, having been officially recognized by the pope in the previous decade. Born of the Counter-Reformation, the black-robed Jesuits had a three-fold mission: to fight Protestant heresy, to spread the gospel, and to educate.

In a land with few Protestants, the Jesuits in Brazil focused on evangelization and education. They created the first schools and, as they would in the rest of the Catholic world, they quickly became the educational and intellectual elite of the Church. The Indian population presented the Jesuits with a formidable evangelical challenge. An extraordinary group of priests, led by Manoel da Nóbrega and José de Anchieta, eventually developed a system of creating Indian villages (*aldeias*) to facilitate Christianization. By the 1560s and 1570s, the Jesuits had gathered together thousands of Indians in dozens of *aldeias*. They taught the Indians the rudiments of Christianity, trained them in European agricultural techniques, and became their defenders before the crown and against the colonists.

In the 1560s, epidemics (most likely smallpox) swept through Indian villages with catastrophic consequences. Starving and sickly Indians staggered into the colonial settlements begging for food and work. With the breakdown of barter, the difficulties of fighting both Indians and Jesuits, and the rising demand for labor on the sugar plantations, the Portuguese began to look for an alternative to native labor. Once again, the Portuguese in Brazil looked to the model of the Atlantic islands and turned to the African slave trade to satisfy their rapidly increasing labor needs. Although the use of Indian labor and enslavement would continue well into the seventeenth century in the core captaincies, by the 1570s the Portuguese had begun a transition to African slave labor that would transform the tiny coastal settlements into a rich colonial enterprise.

SUGAR AND SLAVERY

The Portuguese initiated the Atlantic slave trade in the 1440s, bringing black Africans back to Europe. Slavery dated back to ancient times in Europe and elsewhere, but the enslavement of black Africans by Europeans was new. In 1500, black slaves comprised perhaps 10 percent of Lisbon's population of 100,000. The rise of the transatlantic slave trade, however, signalled the beginning of a long and tragic chapter in world history. For three centuries (roughly 1550 to 1850), Europeans transported their captive human cargo from Africa to the Americas (the so-called "middle passage"). Some 10-12 million Africans survived this forced passage, some 3-4 million going to Brazil alone. Another 5-6 million supplied labor for the mines and plantations of the Caribbean basin. The plantations of the southern United States swallowed up about 750,000 Africans.

In each of these regions, the slave trade created enormous populations of Africans (and their African American descendants), who were dominated by a free white minority. Today, most of the peoples of the Caribbean are the descendants of African slaves. Although the concentration of African Americans in the southern United States has been somewhat altered by migration to large northern and western cities (blacks now make up about 20 percent of the population of the Deep South), African Americans comprise only 11 percent of the U.S. population. In Brazil, some 65 million people (about 45 percent of the country) have African ancestors.

This enormous and centuries-long transatlantic exchange left a profound imprint on Brazilian society and culture. The old plantation regions of the Northeast, especially Bahia, retain striking influences from Africa. Some 80 percent of the people of Salvador are descendants of Africans. Although racial mixture took place throughout the hemisphere, in Anglo America whites refused to accept racially mixed children in their world. Anglo Americans rarely acknowledged miscegenation, and when it occurred, they segregated racially mixed children with blacks in a bipolar society that recognized no middle ground between black and white.

In Portuguese America, a different history unfolded. The shortage of white women, the weak and subordinate position of female slaves, an overwhelming majority of blacks, and a less rigid culture quickly produced a large racially mixed population. Along the coastal zones of the Northeast in the captaincies of Bahia and Pernambuco, the slave trade created a black majority. Mixing between blacks and whites gradually produced a large mulatto population as well. In the interior, the mixing of Indians and Portuguese added to the racial spectrum. These *caboclos* would eventually play a large role in the interior and in the Southeast, especially in the captaincy of São Vicente.

Rather than seeing society in stark, bipolar racial terms, the Portuguese saw color as a continuum and developed a rich and complex vocabulary for describing racial heritage. The racially mixed mulattos and *caboclos* forged a bridge between the world of slavery and the dominant white elite. Cultural mixture also inevitably accompanied racial miscegenation. Roman Catholicism intermingled with African and Indian religious practices. Although Catholicism gradually became the dominant cultural ethos—the "glue" that held Brazilian society together in the eyes of one famous Brazilian social scientist—it was a form of Catholicism deeply infused with non-European symbols and meanings among the non-white masses.

Sugar and slavery made Brazil the first great plantation society in the Americas. Yet, despite the agricultural expansion and the influx of colonists, the Portuguese did not move into the interior of Brazil in any significant numbers. Instead, they clung "crablike to the beaches," as one Brazilian historian put it. Most of those penetrating into the deep interior of the Amazon or the southern captaincies went in search of gold or Indian slaves. These tough adventurers probed the interior and helped Portugal bolster its claims to the vast lands between the coastal settlements and Spanish colonies in the Andes and the La Plata basin. Portuguese presence in the interior, however, was sparse to nonexistent. Three-quarters of the 50,000 Portuguese colonists lived near Salvador and Olinda, the capitals of the sugar captaincies of Bahia and Pernambuco. For every white colonist there may have been as many as three African slaves (not to mention several hundred thousand Amerindians in the interior). Already in the early seventeenth century, the implantation and expansion of sugar had put in place fundamental patterns that continue to plague Brazil: a small white elite controlling vast landholdings, and dominating an economy and political system with a non-white majority.

European imperial expansion in the sixteenth century created Brazil, and in the seventeenth century, European power politics continued to shape Brazilian history. In 1580, after the death of the Portuguese King Sebastian in an ill-fated holy war in North Africa, King Philip II of Spain placed himself on the Portuguese throne through bribery and the threat of force. Philip's mother was Portuguese, and he did have a strong claim to the crown, but many Portuguese resented the move. In history texts, the Portuguese refer to this period (from 1580 to 1640) as the "Babylonian Captivity." The merging of the Portuguese and Spanish empires created the largest empire in world history (only to be eclipsed later by the British Empire). It included control of most of the Americas, the Philippines, the Portuguese trading empire in Asia and Africa, and Spanish possessions across Europe (the Netherlands, Sicily, and southern Italy).

Unfortunately for the Portuguese, the forced coalition with Spain drew them into some bitter European power struggles. Since the fourteenth century, the Portuguese had developed close bonds with

England through dynastic marriages and commercial ties. By the late seventeenth century, Philip II of Spain and Elizabeth I of England had become bitter enemies. More importantly, Dutch Protestants had begun a long war to gain their independence from Catholic Spain. Through banking, trade, fishing, and shipping, by 1600 the Dutch had developed the most dynamic economy in the world. They dominated shipping in the Atlantic, and they carried the bulk of Brazilian sugar, marketing it in Europe. When warfare between the Dutch and Spanish intensified in the 1620s, the Portuguese paid a very high price.

Located deep in central Mexico and high in the Andes, the rich Spanish imperial possessions did not present easy targets for Dutch naval forces. The far-flung commercial and maritime empire of Portugal did. By 1650, the Dutch (and, to a lesser extent, the English) had taken the Asian spice trade from the Portuguese and controlled the Indian Ocean. In Africa, Dutch attackers captured Angolan and West African slave ports and held them for decades. In the 1620s, the Dutch attacked Rio de Janeiro, Salvador, and Recife. After a bloody struggle, they were driven back from Rio and Salvador, but they occupied Recife until the 1650s (although they failed to gain effective control over the sugar hinterland).

In the short term, this global warfare disrupted the highly lucrative Brazilian sugar trade. In the long term, it spelled the end of Brazil's near monopoly of the sugar market. After 1650, the Dutch, English, and French established plantation societies on islands they wrested from Spain in the Caribbean and on the northeastern coast of South America (the Guianas). They soon presented a formidable challenge to Brazilian sugar producers. Barbados, Haiti, and Cuba would replace Brazil and become the dominant sugar exporters in the following centuries. Sugar remained the lifeblood of the colony, but the new competition sent Brazil's economy into decades of decline.

A GOLDEN AGE

The discovery of gold in the interior dramatically altered Brazil's fortunes in the 1690s and helped shift the course of world history. The Portuguese had searched for precious metals in the interior for nearly two hundred years before they hit pay dirt. In the late seventeenth century, Brazilian explorers known as *bandeirantes* began to find gold in the mountain streams to the north of Rio de Janeiro. Often of mixed Indian and Portuguese heritage, these tough *caboclos* probed north and west from frontier settlements at São Vicente and nearby São Paulo. They repeatedly clashed with the Jesuits in their expeditions in search of Indian slaves. One impressed Jesuit remarked that, "They go without God, without food, naked as savages, and subject to all the persecutions and miseries in the world. Men venture for two or three hundred leagues into the *sertão* [interior], serving the devil with such amazing martyrdom, in order to trade or steal slaves."

After 1600, as the African slave trade began to replace the unreliable Indian slave system, the *bandeirantes* increasingly turned to the search for precious metals. By the last decade of the century, they began to find gold in the mountain streams some two hundred miles north of the small port of Rio de Janeiro. Word of the discoveries slowly filtered back to the coast and to Lisbon. By 1700, the Western world's first great gold rush had begun, foreshadowing events in California, Alaska, Australia, and South Africa in the nineteenth century. Thousands of colonists and slaves poured into the rugged mountains north of Rio de Janeiro. The rush eventually spread on a smaller scale to the west (present-day Goiás and Mato Grosso) and received new stimulus with the discovery of diamonds in the region north of the goldfields in the 1720s.

Unlike the concentrated ore in the deep-shaft silver mines of Spanish Peru and Mexico, Brazilian gold was dispersed in the streams and mountains of the Southeast, which made control and taxation of mines and miners very difficult for royal authorities. Gold and diamond production rose dramatically until 1760, which made the Portuguese monarchy the richest in Europe. Sugar, and now gold and diamonds, had definitively established Brazil as the economic heartland of the battered and reduced Portuguese empire. As a popular Brazilian saying goes, "Sugar made Brazil, but gold made it better."

After the Portuguese regained their independence from Spain in 1640, England resumed its traditional role as Portugal's dominant commercial and political ally. A weakened Portugal relied on the British navy to protect it from the French, Dutch, and Spanish. As the English began to manufacture and export more goods (textiles and metal goods in particular) in the eighteenth century, Portugal ran constant trade deficits with its ally. Exporting port wine and cork could not pay for growing imports of English goods, and the Portuguese paid the difference in Brazilian bullion. In effect, Lisbon became a way station for Brazilian gold in the eighteenth century. Ironically, the vast wealth of Brazil reinforced the flaws of a backward Portuguese economy and played a key role in financing England's Industrial Revolution in the late eighteenth century.

In Brazil, the gold rush produced dramatic shifts as well. For the first time, the Portuguese established effective colonization in the interior as the area of Minas Gerais (General Mines) became the most populous in Brazil. The Treaty of Madrid, signed by Spain and Portugal in 1750, moved the old Tordesillas line westward to reflect the lands effectively occupied by the two colonial powers in South America. That line roughly follows the present boundaries of Brazil. The *bandeirantes* and prospectors had extended the reach of Portugal far into the interior, creating a Brazil of continental dimensions.

The flow of goods and people into the Southeast also drained an already weak northeastern plantation economy. In 1763, the king moved the colonial capital from Salvador to the booming Rio de Janeiro, which served as the main entry and exit point for colonists, slaves, and goods in Minas Gerais. This shift of political power reflected the shift in economic and social power that had taken place over the previous fifty years. Power and wealth flowed out of the Northeast to the Southeast, a pattern that has continued for three centuries. Northeastern Brazil continues to pay the price for the political weaknesses of seventeenth-century Portugal, and the geological fortuities of nature.

We will never know how much gold Brazil produced in the eighteenth century. Widespread contraband and smuggling make official production figures highly suspect. Best estimates place gold production at about one million kilos, three-quarters of that coming out of the ground prior to 1770. Minas Gerais accounted for about three-quarters of all production, and the great wealth generated Brazil's first great artistic and architectural boom. Dozens of baroque churches and hundreds of statues and paintings (principally in Minas Gerais) testify to the grandeur that arose from the gold rush. The gold that flowed out of Brazil (principally to England) also fueled the economic expansion of Europe in the eighteenth century. Possibly 80 percent of the gold circulating in eighteenth-century Europe came from Brazilian soil. The gold rush made men rich and powerful on both sides of the Atlantic.

THE AGE OF REFORM AND REVOLUTION

By 1700, a clear shift had taken place in European politics. England and France had replaced Portugal and Spain as the principal European and imperial powers. The Iberians had been at the cutting edge of political and economic change in the sixteenth century. In the seventeenth century, the economic and political innovations of the Dutch, English, and French had carried them to the heights of power in Europe while Spain and Portugal remained tied to the past. Most notably, in the Netherlands and England a new entrepreneurial and commercial class had arisen accompanied by increasing pluralism in the political systems. These innovations fed Dutch and English expansion and in the eighteenth century made England the most powerful nation on earth.

In both Spain and Portugal, the wealth from the Americas made the absolutist monarchies more powerful and less flexible. The continuing power of a traditional landed nobility checked the political and economic pretensions of a very small entrepreneurial and commercial class, thus producing economic stagnation and political inflexibility. With the continuing decline of both empires in the eighteenth century, each made last gasp efforts to reform and renovate. The reforms of the Bourbon dynasty in Spain did open up the economy somewhat, but they could not begin to match the dynamism of England and the Netherlands, where a less-powerful state gave entrepreneurs greater leeway for innovation and risk-taking. Political reform meant the reassertion of imperial control over the Spanish American colonies after a century of relative neglect. This angered the well-entrenched colonial elites, pushed many of them from power, and eventually helped lead to the wars for independence in the early nineteenth century.

In Portugal, the authoritarian Marquis of Pombal (1750-77) attempted similar reforms. This tough prime minister of the weak King José I also made efforts to modernize the imperial system, especially to counter the increasing British domination of the Portuguese economy. Although foreign merchants generally could not participate in the empire's trade, British merchants dominated commerce in Lisbon, and thereby, in the empire. Pombal attempted to "renationalize" imperial commerce by promoting new crops in Brazil and new industries in Portugal, and by improving the balance of trade with England.

In 1789, the reassertion of imperial control and the imposition of new taxes sparked an abortive revolt by colonial elites in Vila Rica, the capital of Minas Gerais. An early sign of Brazilian nationalism, the Minas Conspiracy (Inconfidência Mineira) involved very prominent elite figures as well as military officers. Treason is not a crime treated lightly by absolute monarchs, and royal tribunals sentenced most of the conspirators to prison or exile. The only non-aristocratic member of the conspiracy, a military officer by the name of Joaquim José da Silva Xavier, became the scapegoat. Best known by his nickname, Tiradentes ("Toothpuller"), he was hung, and then drawn and quartered, in 1793. The Crown placed parts of his body on pikes on the road leading into Vila Rica as a warning to those who might contemplate challenging royal authority in the future.

Wisely, Lisbon recognized the roots of colonial discontent and employed persuasion along with power to coopt as well as crush challenges to the imperial system. Over the next few decades (in contrast to the Spanish and British empires), the colonial elites and government in Lisbon worked to strengthen their interdependence. Without Brazilian gold and sugar, Portugal faced economic ruin. Without the support of Portuguese troops, Brazilian miners and planters faced the specter of rebellion by the slave majority. In 1791, slaves rose up in Haiti and drove the French from the island after years of bloodshed. The Haitian Revolution (the only successful slave rebellion in the Americas) gave white planter minorities nightmares throughout the Americas. Brazilian slave owners, living amidst a slave majority, understood the fragile repressive line between order and chaos, and they were not overly anxious to challenge established authority.

Portuguese (and Spanish) colonists, as a rule, rarely chose to challenge the system. On the contrary, Iberian immigrants sought success by rising through the hierarchical bureaucracy of the Spanish and Portuguese monarchies. Their religious and political preferences reinforced the power of king and church rather than challenging or questioning them. Although the so-called Thirteen Colonies shared many of the same problems as the Portuguese empire in Brazil (and the Spanish American colonies), they had already begun to follow a distinct historical path by the late eighteenth century. Many of the English colonists had left Britain fleeing religious and political persecution. Even those who had not fled, already shared the values of a political culture that since the late Middle Ages had increasingly placed checks on royal power and recognized individual rights.

In all the American colonial societies, the efforts by the monarchies to impose taxation without significant political representation produced revolts among the colonists. In the Thirteen Colonies in the 1780s, the war for independence succeeded. In the Iberian colonies the wars for independence did not come until the 1810s and 1820s. With little to exploit in either agriculture or raw materials, the British colonies in New England developed a substantial population of small farmers and merchants, as well as a vigorous shipping trade with Europe and the Caribbean. After dominating or driving back the relatively sparse Indian peoples of eastern North America, the southern British colonies looked very similar to the core regions of Latin America—plantation agriculture, a large unfree labor force, and an elite white minority. The commercial dynamism of New England, and the lack of a plantation past, enabled it to become a vibrant economic center that emerged by the early nineteenth century as the engine of the new country's economy. The new states of the southern United States, and the Spanish and Portuguese colonies, however, would enter the nineteenth century burdened with the social and political inequities of plantation economies.

THE PATH TO INDEPENDENCE

By the nineteenth century, well-entrenched and sophisticated colonial elites had emerged in many parts of Latin America. In Brazil, the strongest and most mature of these elites resided in the Northeast (Salvador and Recife) and in the Southeast (Minas Gerais and Rio de Janeiro). Smaller, but substantial, regional elites had developed in a number of other coastal centers, most notably in the North (Belém and São Luís) and the South (the captaincy of Rio Grande do Sul). These elites increasingly sought greater economic autonomy and political representation in Portugal. The Enlightenment, the American and French Revolutions, and the growing economic power of England's Industrial Revolution deeply impressed the Brazilian elites. The sons of these elites who sought a university education had no choice but to cross the Atlantic, at which point they experienced firsthand the seismic economic, intellectual, and political shifts that were reshaping Europe.

While the discontent of colonial elites had grown dramatically by the beginning of the nineteenth century, unlike their U.S. counterparts, these elites had not forced the issues of political and economic autonomy to the breaking point. War in Europe broke the colonial bonds for them. In 1807 and 1808, Napoleon severed the connection for the colonies when his armies invaded Portugal and Spain. In the first decade of the nineteenth century, Napoleon dominated Europe from Russia to the Atlantic. Unable to challenge British control of the seas, he turned to the conquest of the last two continental European regions not under his control: Iberia and Russia. Although he quickly occupied Spain and Portugal, continual warfare in both countries sapped his armies for years. The conquest of Iberia did allow Napoleon to turn in 1812 to the invasion of Russia, where he would face defeat and the collapse of his European empire.

Napoleon's control of Spain, and Britain's control of the seas, left the Spanish American colonies adrift and set off a chain reaction that led to the wars for independence. By 1826, all of Spain's colonies in the Americas (except Cuba and Puerto Rico) had achieved independence. Unlike Britain's North American colonies, which were a minor outpost of a world empire, the Spanish American colonies formed the core of Spain's once-mighty empire. Spain fought fiercely and futilely to retain what it could, with the bloodiest wars taking place in the richest colonial centers: Mexico and Peru.

In Portugal and Brazil, the historical path took a decidedly different turn. For at least a decade, the Portuguese monarchy had anticipated a French invasion, and when it came in 1807, the crown did not accept the surrender and imprisonment that would be the fate of the king of Spain. Recognizing that Brazil *was* the Portuguese economy, and preferring retreat to imprisonment, the Portuguese monarchy fled Lisbon shortly before French troops entered the city. Ten thousand Portuguese joined the royal family on British ships in November 1807 for an unprecedented voyage across the Atlantic. With the help of their British allies (Napoleon's bitter enemies), the Portuguese monarchy transferred the center of the empire to Rio de Janeiro. For the first—and last—time in Western history, a European monarch would rule his empire from the colonies.

Prince Regent (and later King) João arrived in Brazil in early 1808 and for the next thirteen years ruled Portugal's Asian, African, and American colonies from the court he constructed in Rio de Janeiro. While the Spanish American colonies warred with Spain for their independence, Brazil flourished as the center of the Portuguese empire. João established the cultural and political institutions of an imperial center, institutions that Brazil had sorely lacked. By 1821, 150,000 of Brazil's 3 million inhabitants lived in Rio. (In comparison, New York City had 125,000 inhabitants in 1820 when the United States had a population of about 9 million.) Slaves probably comprised half the colony's population, the racially mixed accounted for another quarter of the inhabitants, and the Portuguese-born (known as *mazombos*) probably numbered about 100,000. In 1815, João VI elevated Brazil to the status of a kingdom, placing it on an equal footing with Portugal. The presence of the monarchy and court in Rio brought Brazilian and Portuguese elites together, and it paved the way for a gradual transition to independence.

The end of the Napoleonic wars in Europe in 1815 opened the way for the monarchy to return to Lisbon, but João resisted and remained in Brazil. In 1821, a new and aggressive Portuguese parliament (the Cortes) produced a constitution that restricted the king's power, and also returned Brazil to colonial status. Threatened with the loss of his crown, João reluctantly returned to a Portugal split by political intrigue. Legend has it he left his twenty-three-year-old son, Pedro, in Brazil with some sage advice. João recognized the desire of Brazilians for self-rule and saw that the Cortes wanted to return to the old imperial system. Wishing to avoid the bloodshed that had fragmented the Spanish American colonies, he warned Pedro not to fight the rising movement for independence. Instead, he told him to join and lead the movement if it became powerful. The king, in effect, told the crown prince to rebel against the monarchy in the event that conflict emerged. Better to have father and son on two thrones than to lose Brazil to revolutionary leaders.

Pedro followed his father's advice. His refusal to return to Portugal, and his defiance of orders from the Cortes, cemented his role as the leader of independence. On September 7, 1822, while traveling in the interior near São Paulo, Pedro stopped by a small stream (the Ipiranga) for a brief rest. A messenger arrived with letters from the Cortes that once again challenged his authority. With this came a letter from his closest Brazilian advisor urging Pedro to seize the moment and to break with Portugal. According to one witness, Pedro threw down the letter from the Cortes, ground it under his heel, and drew his saber. With a flourish, he waved the sword and declared, "Independence or death! We have separated from Portugal!" The "Cry of Ipiranga" has been celebrated by Brazilians ever since as their independence day.

England acted as a midwife in the birth of this South American nation. The English had long dominated Portugal's economy and its foreign policy, and the split between crown and colony left the British government in a difficult but pivotal role. Wanting to protect its interests on both sides of the Atlantic, Britain handled negotiations between Lisbon and Rio de Janeiro. With few troops in Brazil, and civil war erupting between absolutists and constitutionalists at home, Portugal could do little to counter Pedro's unilateral declaration. Pedro secretly agreed to pay Portugal 2 million pounds sterling (roughly $10 million) in compensation for royal properties in Brazil. He also made some formal public concessions in exchange for official Portuguese recognition of Brazil's independence. For their part, the British established themselves (through treaties) as Brazil's dominant trading partner.

While the divorce from Portugal may have been relatively painless, the struggle to keep the family together in Brazil was not. Pedro's greatest challenge was to keep this new nation of continental dimensions from fragmenting into several countries as had happened in Spanish America. For the next two decades, regional revolts threatened to shatter national unity. Many of these revolts revealed deep and bitter social and racial cleavages in the new nation. Several of these revolts embodied the most serious challenges by the masses to elite rule since the founding of the colony. Lord Cochrane, a mercenary admiral who had been thrown out of the British navy, performed small miracles by crushing revolts in the major regional centers (all of them along the coast). Thus, with a small hired navy and in very few battles, Brazil gained its independence in the 1820s.

Pedro's leading role in the path to independence, however, rendered it incomplete. Pedro I, although in Brazil since the age of ten, was Portuguese. With a liberal heart and an authoritarian mind, he ruled the Empire of Brazil using a constitution written under his supervision and promulgated against the opposition of much of the Brazilian elite who had supported his earlier moves. Pedro's constitution created a centralized regime with a weak parliament and a strong emperor. Although the new regime spelled the end of an absolute monarchy, the foundations of a government with limited powers set out in the Constitution of 1824 were ideals rather than reality. The idea of government as a contractual agreement between people and ruler had barely begun to penetrate a political culture that was built on rule by inheritance. Brazilians had traded a Portuguese monarch in Portugal for a Portuguese monarch in Brazil, albeit one with liberal sentiments.

The emperor's autocratic tendencies, a weak economy, war, and Pedro's complicated personal life made Brazilians increasingly dissatisfied with their Portuguese ruler during the decade after independence. Pedro had a disturbing tendency to override the parliament and surround himself with Portuguese-born cabinet ministers. Pedro's huge payment to Portugal did not remain a secret for long, and it provoked the anger of elites who blamed the Emperor for the national economy's poor performance. In the late 1820s, Pedro also unwisely chose to renew a long-standing struggle with Argentina over the southern border of Brazil. The Cisplatine War (1825-28) drained vital and scarce financial resources from the government. (England mediated a settlement that created the nation of Uruguay as a buffer state between South America's two most powerful nations.)

In addition, a highly visible and prolific extramarital love life rocked Pedro's two marriages and further complicated his relations with the Brazilian elite. Although discreet affairs might have been accepted and pardoned in this patriarchal, macho culture, discretion was not one of Pedro's strong points. For much of the 1820s, he carried on a very public and torrid love affair with Domitila de Castro, the wife of an army officer. She bore his children, and sometimes accompanied him on official state functions, and he eventually ennobled her as the Marchioness of Santos. Although the marchioness was well liked by the Brazilian elite, Pedro's blatant disregard for his wife, Leopoldina, and their bitter arguments, created a great deal of unnecessary ill will.

Finally, Pedro remained deeply involved in the affairs of his native Portugal. His father died in 1826, leaving Pedro heir to the Portuguese throne. Faced with rebellion by his Brazilian subjects in 1831, he decided to abdicate and return to Portugal. He would fight absolutists to place his daughter on the throne as a monarch with strong constitutional restraints. Ironies abound in this turn of events.

In response to the liberal constitutionalists in the Portuguese parliament, Pedro had declared Brazilian independence. In Brazil, he faced political foes who rejected his own autocratic style. Now he returned to fight his brother, Miguel, and the forces of absolutism to replace them with a constitutionally restricted monarchy. For all his flaws, Pedro served Brazil well in two crucial moments. In 1822, he helped move the country into nationhood with a minimum of bloodshed, and in 1831, he again prevented war by abdicating his empire instead of fighting back. As one astute French observer on the scene noted of Pedro's tumultuous ten-year reign and abdication, "He knew better how to abdicate than to rule."

Like his father, Pedro left behind his eldest son (also named Pedro) to take his place. The child of Pedro I and his Austrian wife, Leopoldina, the future Pedro II would become the dominant figure in nineteenth-century Brazil. After his fifth birthday, in 1831, the year his father and family returned to Portugal, the young Pedro grew up a virtual orphan and received an extraordinary education. Carefully selected tutors taught the future emperor Latin, Greek, French, German, Spanish, and English, and gave him a broad and profound education in the arts and sciences. Throughout his long life, Pedro would remain an intellectual with a powerful curiosity. In many ways, Pedro became the enlightened monarch envisioned by the philosophers of the eighteenth century.

While the young emperor-to-be grew up, a council of regents appointed by the parliament ruled the country. For the first time, Brazilians governed Brazil. As in most of nineteenth-century Latin America, two political parties contended for power. Conservatives looked back to Iberian values and traditions for their inspiration and sought to maintain the influence of the Catholic Church, a strong centralized monarchy, and the slave economy. Liberals sought to mold their country in the image of England, France, and the United States. They wanted to diminish the influence of the Church, restrain central-ization and monarchy, and move toward a free labor economy. These were the ideals. When in power, each tended to be very pragmatic, sometimes implementing their opponents' programs.

Throughout the 1830s, the absence of a strong executive, disputes between Liberals and Conservatives, and powerful regional revolts threatened to shatter the tenuous unity of the new nation. The constitution did not allow for the coronation of young Pedro until his eighteenth birthday in December 1843. The exceptional maturity of the heir, the desires of both parties for the stability provided by a monarch, the everpresent fear of slave rebellion prompted by the numerous regional revolts, and the hopes of both parties that they might dominate the teenager, all converged in 1840, bringing Pedro to power at the age of fourteen. In what amounted to a legal coup d'etat, the parliament offered Pedro the crown—and he accepted—beginning a reign of forty-nine years known as the Second Reign (1840-89).

A BRAZILIAN EMPIRE

Despite the declaration of political independence in 1822, the 1840s mark the true transition from colonial to modern Brazil. In politics, the gradual transition begun with the transfer of the court in 1808, independence in 1822, and abdication in 1831, comes full circle with the coronation of Pedro II in 1841. After twenty years of independence, a Brazilian-born monarch finally ruled over the regime put into place in the 1820s. The government definitively crushed the last of the major regional revolts in the 1840s, thus consolidating nationhood. By mid-century, the central government and the economic and political elites of the coast had achieved control (even if somewhat tenuous) over most of the country. Finally, the 1840s also mark the emergence of coffee cultivation as the engine of economic growth that would transform Brazil for the next century.

Like sugar, coffee is not indigenous to the Americas. In the eighteenth century, the Portuguese planted strains of Middle Eastern coffee in northern Brazil, and in the following decades cultivation spread, reaching the fertile valleys near Rio de Janeiro in the 1820s and 1830s. During the next century, coffee cultivation spread rapidly in the area north and west of Rio, in southern Minas Gerais, and (most prominently) in the province of São Paulo. The rapid expansion of coffee fields quickly made Brazil the world's leading exporter, a position it continues to hold today. Revenue generated by the coffee economy drove the Brazilian economy until the Great Depression in the 1930s and definitively established southeastern Brazil (principally the states of Rio de Janeiro, Minas Gerais, and São Paulo) as the economic and political core of the nation.

Export taxes on coffee provided the vast bulk of government revenue for expanding the bureaucracy, and for building roads, ports, and communications systems. Coffee exports allowed Brazil to maintain a favorable trade balance throughout the Second Reign and beyond. By the 1920s, the expansion of the coffee economy in São Paulo made the city and the state into Brazil's most powerful economic and political force. The dominance of sugar in the Cuban economy has often been summed up in the saying, "Without sugar, there is no country." In the case of nineteenth-century Brazil, one could say, "Without coffee, there would have been no modern Brazil." Coffee served as the catalyst in the transformation of nineteenth-century Brazil.

As did sugar in the seventeenth century and gold in the eighteenth century, the coffee economy ran on slave labor. Brazil imported half a million slaves in the seventeenth century to toil on the sugar plantations of the Northeast. In the eighteenth century, the goldfields of Minas Gerais absorbed another 1.5 million Africans. In the first half of the nineteenth century alone, Brazil imported another 1.5 million slaves to fill the demand for labor on the coffee plantations of the Southeast. As the abolitionist movement gained strength in England and the United States in the nineteenth century, British naval pressure forced Brazil to halt its 300-year-old Atlantic slave trade in 1850.

The entry of 3.5 million Africans into Brazil in a process that had accelerated in volume right up until 1850 fundamentally shaped the composition of Brazilian society. In 1800, Brazil had the largest slave population in the world (half of its population of 3 million) and this forced migration created a truly African American culture in Brazil. African music, religions, foods, and language patterns blended with the culture of the Portuguese and the Indians to produce a cultural mosaic that was not African, European, or Native American.

For nearly four decades after the end of the Atlantic slave trade, the Brazilian elites debated and legislated, slowly chipping away at the slave system. Slavery had been so central to the fabric of life in Brazil for so long that dismantling it took much longer than in any other society in the Americas. While personally opposed to slavery (he freed all the slaves he inherited), the emperor did not force the issue. With the rise of a vocal abolitionist movement in the 1880s (largely in the cities), and the growing tendency for slaves to flee from their masters, the system began to disintegrate. Legislation by the Conservatives attempted to stretch the process over decades by freeing the children of slaves gradually (after 1871) and by emancipating elderly slaves (after 1885). By 1888, unrest on plantations and the refusal of the army to step in and halt the flight of slaves from their masters brought the system to the brink of chaos. Ruling in place of her father, who was in Europe for medical treatment, Princess Isabel decreed the end of slavery with the "Golden Law" of May 13, 1888. Rather than face the anarchy and upheaval of massive slave unrest and flight, slaveowners grudgingly accepted abolition. Pressure from both above and below doomed Brazilian slavery.

A flood of European immigrants to Brazil also eased the process of abolition. With the supply of new captive labor cut off after 1850, coffee planters turned to European immigration to meet their labor needs. The pursuit of European labor coincided with the "Great Migration" of peoples out of Europe in the late nineteenth century. Some 60 million Europeans left their homelands between 1815 and 1930 in search of a better life. The vast majority crossed the North Atlantic and settled in the United States. A smaller, but significant, migratory wave populated Argentina and Uruguay with European peoples. Some 2.7 million immigrants—mainly from Italy, Spain, and Portugal—arrived in southeastern and southern Brazil between 1887 and 1914. By the turn of the century, the majority of the citizens of the city of São Paulo were immigrants or their children. These immigrants gradually replaced slaves as the labor force in the coffee fields, and they turned southern Brazil into a branch of European civilization strikingly different from the older mining and plantation regions of Minas Gerais and the Northeast.

Despite the fundamental shifts that were gradually transforming the economy and society, Brazil maintained an enviable political stability during the Second Reign. In stark contrast to the upheaval and instability of countries like Mexico and Peru, Brazil developed a power-sharing arrangement between Liberals and Conservatives. The emperor acted as a "moderating power," almost a fourth branch of government, calling for new elections when it appeared that the ruling party faced a political crisis. Invariably, the opposition party would win the new (and highly restricted) elections. In forty-nine years, power shifted hands between the two parties twenty-six times, a remarkable feat of conciliation and political consensus, even for nineteenth-century Europe. Pedro II reigned over what some viewed as a tropical version of Queen Victoria's parliamentary regime in England.

Although the political elite did not act as a monolithic bloc, they did exhibit extraordinary cooperation. Both parties accepted the monarchy and the emperor's role. In the 1870s and 1880s, a republican movement emerged that called for the end of the monarchy and the creation of a republic modeled after the United States. The Republicans played a prominent role in the struggle for abolition, and they saw the monarchy as another outmoded institution that held back the development of Brazil. The Republicans, and other members of the Brazilian elite, were deeply influenced by the writings of the French philosopher, Auguste Comte (1797-1857), and the English thinker, Herbert Spencer (1820-1904). Known as positivism, Comte's philosophy glorified reason and scientific knowledge and rejected traditional religious beliefs. He believed that humanity had begun to overcome the superstitions and religious beliefs of the past, and, with the guidance of science (positive knowledge), it would soon enter into a new age where technicians and engineers would run an authoritarian republic to achieve true progress. Comte summed up his beliefs in a catchy epigram: "Love as the principle. Order as the base. Progress as the goal." In Spencer, Brazilian elites found a sophisticated scheme of social evolution that not only glorified science and reason, but also provided them with a positivistic rationale to justify their control of the "inferior" racially mixed lower classes.

While the writings of Comte and Spencer inspired many Brazilians, they exerted a diffuse but especially powerful influence on the army. Several influential instructors in the military academies returned from France with technical degrees and a passion for positivism. Beginning in the 1870s, these positivists imbued generations of military officers with the authoritarian, scientific, and industrial vision of Comte and Spencer. The influence of positivism would prove crucial in the fall of the empire, and in the role of the military in Brazilian history for the next century.

During the twentieth century, the military has played a central (and sometimes *the* central) role in Brazilian society. Brazil avoided most of the bloodshed and huge military buildup that in the early years plagued the new Latin American nations. Pedro II, fully conscious of the problems of his Latin American neighbors, strove to keep the military weak and underdeveloped prior to the 1860s. An international war began the transformation of the Brazilian military. For complex reasons, Brazil joined Argentina and Uruguay in a long and costly war against Paraguay in the 1860s. Despite an enormous disparity in resources, the Paraguayans tenaciously resisted the invading armies for six bitter years (1864-70), losing the majority of its adult male population and large chunks of territory during the conflict.

Brazil's inability to defeat tiny Paraguay quickly and effectively highlighted the military's weaknesses. Pedro grudgingly allowed the buildup for the war effort, and his efforts to scale down the army and navy after 1870 alienated and irritated many officers. The emperor reacted swiftly and decisively, squelching efforts by some officers to enter political debate, and thus creating more ill will in the officer corps in the 1870s and 1880s. Positivism and republicanism provided disaffected officers with an authoritarian model that seemed to point to an industrial future for Brazil, minus the monarchy, the church, and slavery.

Conflict between church and state also characterized the 1870s and 1880s. Continuing the Iberian tradition, the church in most of Latin America operated as a virtual arm of the state. In places like Colombia, this resulted in a very powerful institution that was intimately tied into the political power structure. In Brazil, state control of the church strangled the institution. Privately an agnostic, Pedro used his power as emperor to restrict the development of the church in Brazil. He restricted the number of seminaries, the formation of new priests, and the creation of new dioceses, and he fought with the pope on occasion. His refusal to publish a papal bull, and the imprisonment of a bishop for defying his authority, brought the Catholic Church directly into conflict with the monarchy by the 1880s.

Much like the transition from colony to empire, the shift from monarchy to republic initially transpired with little bloodshed, only to be followed in succeeding years by powerful challenges to the authority of the central government. By 1889, an ailing, sixty-three-year-old Pedro had lost the support of key groups in the imperial power structure, and his heir, Princess Isabel, had few supporters among the elites. Abolition, church-state conflict, republicanism, and positivism had all eroded Pedro's traditional pillars of support among the landowners, clergy, and military. The demise of slavery, and with it the threat of slave rebellion, also removed the long-standing fear that had persuaded landowners to think twice before challenging the monarchy. A small group of conspirators, with key support from high-level army officers, initiated a coup d'etat on November 15, 1889. The surprised Pedro found himself with little support and wisely (like his father) chose exile over resistance. The following day, the royal family sailed to exile in Portugal and France. Within a month, the empress died, and Pedro mourned the loss of wife and country for another two years before dying in a hotel in Paris in 1891. Banned from Brazil, he lay buried in the royal mausoleum in Portugal alongside his father and grandfather. The bodies of Pedro and the empress were returned to their beloved Brazil for reburial in 1922, during the celebration of the centennial of independence.

COFFEE-AND-CREAM POLITICS

Political outsiders dominated the new republican regime. Positivists, republicans, military figures, and the middle class suddenly replaced the traditional political elite, which was dominated by landowners. Until 1894, the military ran Brazil's First Republic, initially under the leadership of Deodoro da Fonseca, a conservative general who had joined the revolt at the last minute. A constitutional congress reluctantly named Deodoro president while enthusiastically electing another, more decidedly republican general, Floriano Peixoto, as his vice-president. A new constitution (using the U.S. charter as a model) went into effect in 1891, replacing the imperial constitution that had served the nation for sixty-seven years. A new flag, bearing the slogan "Order and Progress" (Ordem e Progresso), reflected the influence of positivism among the leaders of the new regime.

Deodoro feuded constantly with the new congress over legislative matters and angrily resigned the presidency in 1891. The tough Floriano assumed control and guided the republic through difficult times. Refusing to tolerate challenges, the "Iron Marshal" did not shrink from using firing squads, imprisonment, and the forced retirement of generals to consolidate the power of the new republic. A revolt by the navy nearly toppled the regime, and the world economic depression of 1893-94 shook the coffee-export economy. With the fate of the new republic in the balance, the *paulista* (São Paulo) planters and politicians strongly supported the central government. Their help was rewarded when

Floriano supervised elections in 1894 and handed power over to a civilian president, Prudente de Morais Barros, a *paulista,* ending the "Republic of the Marshals." In doing so, Floriano established a pattern that the Brazilian military would follow many times in the coming decades. They had stepped into the political crisis in 1889, acted decisively to ease the transition from the old regime to a new one, and then stepped back from power. The birth of the republic marked the emergence of the military as the key power broker in national politics.

With the election of Prudente, the powerful coffee interests returned to dominance in national politics. Over the next four decades, the coffee states of São Paulo, Rio de Janeiro, and Minas Gerais shared political power and the presidency. Nine of the twelve presidents between 1894 and 1930 came from these three states. The southernmost state of Rio Grande do Sul often joined the coalition as a subordinate partner and contributed one president. With Minas best known for its dairy products, and Rio and São Paulo for their coffee, historians have coyly referred to this political coalition as the "coffee-and-cream alliance." The three states produced most of Brazil's wealth and accounted for most of its population.

The First Republic was both highly controlled and decentralized. Less than 1 percent of the population could vote, and the coffee oligarchy ran state and national affairs. To appease the less powerful states, and to guarantee stability in the interior, the federal government struck a political deal. Recognizing that the central government did not have the revenues nor the means to extend its power into the interior, where 80 percent of all Brazilians lived and worked, the political elite decided not to disrupt traditional power relations in the countryside. For centuries, local landowners (honorifically known as colonels of the local militias) dominated the rural population by the control of local government and local courts, and through the use of hired guns. Under the First Republic, these landowners retained their local power in exchange for allegiance to the state government. The state provided the colonel with outside support if needed, and the colonel guaranteed local votes for the "official" state candidates. In turn, the states pledged support to the federal government in exchange for the right to run their own affairs. Should the state officials require help to crush challenges to their authority, the federal government sent troops and supplies to assist the state militia. The federal government

accepted this decentralization, but acted swiftly when state governments failed to demonstrate the necessary allegiance. Federal troops
intervened in political crises in every state in Brazil (save São Paulo) at
some point during the First Republic.

This decentralized system permitted the federal government to
maintain its authority, states to run their own affairs, and local landowners to maintain their traditional powers. Despite the rhetoric of the
republic's leaders, and the hopes of revolutionaries in 1889, the new
regime represented continuity with the past. Although Brazil began to
experience some industrialization, and urbanization accelerated, the
traditional Brazil in the countryside remained largely unchanged. Under
the First Republic, a thin veneer of modernization (largely in the cities)
hid the vast traditional rural society, in which a few landowners
dominated the lives of millions of country people.

Periodically, this traditional rural society erupted, challenging the
emerging modern, urban Brazil. Banditry and revolts in the interior
occasionally reminded politicians in Rio and São Paulo of the "primitive," non-European side of their society. The most spectacular of these
eruptions took place in 1897 at Canudos, a village deep in the rugged
interior of Bahia. A religious mystic known as Antônio the Counselor
gathered thousands of followers around him, denounced the secular
republic, and called for its destruction. After the Counselor and his
followers routed a local police contingent, then two large state militia
expeditions, the federal army stepped in. These poor, rural folk fero-
ciously resisted, holding at bay several thousand federal troops during a
military siege lasting months. The Canudos revolt revealed the powerful
resistance of traditional rural folk to the encroachment of modern
European society and the modern state into their lives. Conselheiro died
during the struggle, and the army ruthlessly obliterated the town and its
inhabitants at the end of the siege.

Prior to World War I, the Republic faced few major challenges
from urban-based groups. The oligarchical regime faced just two
seriously contested elections (1910 and 1922) during four decades. In
each case, disaffected sectors of the elite challenged the coffee-and-cream
alliance, only to face certain defeat in the face of unity among the three
state powers. By the end of World War I, however, new forces had begun
to emerge in modern, urban Brazil that would challenge, and then help
topple, the coffee oligarchy.

By the 1920s, small industry had begun to develop, especially around the cities of Rio de Janeiro and São Paulo, and the growth of a small working class and middle class accompanied this economic development. Both groups found themselves excluded from a power structure developed by landowners to dominate rural workers. The huge numbers of immigrants, particularly Italians, in São Paulo and Rio made up a large percentage of the working class and introduced them to new political ideologies. (In 1920, half of the workforce in the city of São Paulo were foreign-born, and a good deal more were the children of foreigners.) Socialists and anarchists organized unions and strikes, encountering intense repression from the government. Middle-class discontent with the republic was more diffuse and less focused around a particular ideology or movement.

Once again, the military became a focal point for dissatisfaction with the regime. Although the direct influence of Comte had virtually disappeared by the 1920s, positivist ideas (in a broad sense) continued to influence the military, especially junior officers in the military academies. These officers, more than their predecessors in the 1890s, wanted to see an industrializing, technologically advanced Brazil. In their eyes, the coffee oligarchy had diverted the republic from the ideals of its founding fathers, and had held the country back with policies that favored agriculture, especially coffee production. Many of these young lieutenants (*tenentes*) dreamed of a Brazil led by technocrats instead of coffee barons.

In many ways, twentieth-century Brazil did not emerge until the 1920s. A series of cultural and political events in 1922—the centennial of independence—symbolized the emergence of a new Brazil. An avant-garde group of artists, intellectuals, and writers organized a Modern Art Week in February 1922 that announced the renovation of Brazilian culture. Though inspired by the modernist movement in Europe, these intellectuals declared Brazil's cultural independence and denounced centuries of slavish imitation of European culture. The central project of Brazilian intellectuals, in their eyes, was to "rediscover" Brazil and to develop an authentically Brazilian culture, one that recognized not only the nation's European roots, but also the cultural contributions of Africa and America.

The year 1922 also marked the founding of the Communist Party of Brazil (PCB). Although by no means the only, or the most important socialist political movement in the country, the PCB would eventually become a potent political force, and its birth symbolized the emergence of the left in Brazilian politics. The most potent challenge to the regime, however, came not from radical leftists, but from radical military officers. On July 5, 1922, a group of *tenentes,* attempting to topple the government, rose up in an abortive and bloody revolt in Rio de Janeiro. While the revolt failed miserably, their heroic stand on the beach at Copacabana Fort signalled the beginning of serious armed resistance to the republic. On the anniversary of the revolt in 1924, a more potent, weeks-long revolt by *tenentes* in São Paulo shook the foundations of the regime. One large group of rebels (who fought in Rio Grande do Sul before converging on São Paulo) retreated into the interior under the command of Lt. Luís Carlos Prestes, operating as a guerrilla force. For more than two years, the Prestes Column moved through the interior, covering nearly 14,000 miles before disbanding in Bolivia.

By the late 1920s, the challenges of young army officers, middle class groups, urban workers, and intellectuals threatened the stability of the regime but did not have the power to bring it down. A serious rift within the republican elite would. The governor of Minas Gerais, Arthur Bernardes, became president in 1922, and the governor of São Paulo, Washington Luís Pereira de Sousa, succeeded him in 1926. Washington Luís angered and alienated the Minas political elite when he picked another *paulista* politician to succeed him in 1930. Already deeply troubled over the impact of the world depression that had begun in 1929, the Brazilian political elite split asunder. Powerful politicians in Minas combined with the political establishment in Rio Grande do Sul and other states, forming an alliance to challenge the official *paulista* candidate, Júlio Prestes. After a bitter and violent campaign, Prestes "won" the election. The assassination of João Pêssoa, the opposition vice-presidential candidate, galvanized the alliance into armed revolt in October 1930. After a month of fighting, Washington Luís stepped down as rebel troops marched into Rio de Janeiro. The "Revolution of 1930" had triumphed.

GETÚLIO VARGAS AND THE NEW BRAZIL

The civil war in Brazil formed a part of the larger upheaval that arose out of the crisis sweeping through the capitalist world after 1929. The sudden collapse of the Western economy aided the rise of fascism and nazism in Europe, and stimulated the development of social welfare states in Scandinavia, Britain, and the United States. Political leaders throughout the Western world fell from power, and the new leadership reformed or discarded the old politics and economics. In Brazil, the "revolution" did not fundamentally transform politics and society; rather, it produced an important realignment of forces. Rather than a dramatic break with the past, the revolution represented the blossoming of the seeds of unrest that had been planted in the 1920s.

The old politics had been developed over decades to fit a Brazil that no longer existed, a society composed primarily of two important classes: a huge majority of rural workers, and a tiny minority of landowners who controlled the system. By 1930, the emergence of new urban groups—especially industrial workers, industrialists, and the middle class—made the old politics unworkable. The small but increasingly important urban groups did not have the power to wrest control from the coffee oligarchy and their allies. The split within the oligarchy over the 1930 election, however, opened the door for the new groups. In essence, Brazilian politics since 1930 has been the story of efforts to produce a workable and stable realignment of the still-powerful old forces, as well as the increasingly powerful new ones.

Getúlio Vargas played the central and decisive role in these efforts and stands out as the most important political figure in twentieth-century Brazil. A complex and fascinating man, Vargas promoted a politics of social welfare not unlike Franklin Delano Roosevelt's, and (for a time) repressive tactics like those of Francisco Franco in Spain and António Salazar in Portugal. Vargas rose swiftly to prominence in national politics. Born in 1887 into an elite ranching family near the Argentine border, he was a *gaúcho* (a native of Rio Grande do Sul). The son of a general, Vargas eventually graduated with a law degree, after abandoning both the military academy and pharmacy school. In less than a decade (1922-30), he went from being federal deputy to finance minister in the cabinet of Washington Luís, to governor of his home state of Rio Grande do Sul, to presidential candidate and leader of the revolutionary coalition.

Much like FDR, Vargas was a conservative modernizer, a member of the elite who recognized the need to accommodate new political and economic forces into the old system. Although firmly rooted in the old

Brazil, Vargas acted as the midwife for a new Brazil. An extraordinarily adept politician, Vargas understood Brazilians and the national psyche better than any Brazilian politician before or since. He brokered a realignment of Brazilian politics during the next quarter-century through a brilliant juggling act, playing off the diverse interests of social and political groups. From 1930 to 1934, he did this as the head of a provisional revolutionary government (a "Second" Republic), then as the president elected by a constituent assembly in 1934. With the help of the army, he led a coup in 1937 and for the next eight years ruled the nation as the dictator of a "New State" (Estado Novo) with the superficial trappings of fascism (a "Third" Republic).

As World War II came to an end and vast changes again swept across the Western world, the military decided Brazil needed a change as well. Despite Vargas's promises to hold national elections, a suspicious military high command forced him to relinquish power in October 1945. In an impressive demonstration of his popularity, two states elected the deposed dictator as their senator in December 1945, and seven states elected him to represent them in congress. Vargas returned to the presidency (1950-54) elected by a majority of the voters in democratic elections.

Getúlio Vargas helped create a new style of politics known as populism that challenged the "old boy network" of the coffee oligarchy. Brazilian populism after 1930 combined the rhetoric of economic nationalism with social welfare programs. Populists forged a political following among the masses (especially urban workers) and sectors of the middle class. Populism did not challenge the traditional structures of society and politics with revolution; instead, it accommodated the new social groups into the power structure through reforms. Vargas, for example, continued government subsidies for the coffee planters, and he did not attack that most basic and sensitive of issues—land reform. The extremely concentrated structure of land in the hands of a few, and the old problem of millions of landless peasants in the countryside, remained unchanged and unchallenged.

To satisfy his urban base, Vargas stressed the creation of new industries, Brazilian industries, through protectionist policies in the 1930s and 1940s. (Instead of importing manufactured goods from abroad, workers would produce these goods in factories in Brazil.) These policies pleased an emerging new class of entrepreneurs and industrialists and created more jobs for blue-collar and white-collar workers. Finally, Vargas did initiate a social welfare revolution. Much like Roosevelt's New Deal, Vargas' legislation provided workers with basic social welfare protections: minimum wage, maximum working hours, pensions, unemployment compensation, health and safety regulations, and unionization. In a nation just beginning to experience industrialization, these new laws applied to a very small portion of the urban workforce, and represented small but important innovations.

Through planning and the unanticipated effects of government monetary and fiscal policies, Brazilian industry surged after 1930. For the previous half-century, the revenues of the coffee economy had helped finance small-scale industry, especially around the city of São Paulo. By 1930, the city had assumed its role as the engine of the national economy. Economic nationalism and protectionism stimulated the growth of industry and helped Brazilians substitute domestically produced goods for foreign manufactures. In the nineteenth century, Brazilians had imported many of their consumer goods. The elite had imported everything from the soap and toothpaste they washed with to the shirts and shoes they wore. By the 1940s, Brazilian industry produced about 90 percent of all consumer goods. This helped the balance of trade and balance of payments problems of the 1930s, and it further stimulated local industry.

Under Vargas, the government made impressive efforts to promote the development of basic industries. In the 1940s, for example, Vargas led a successful fight to build South America's first integrated iron and steel complex at Volta Redonda (near Rio de Janeiro). Astutely playing on U.S. and British fears of German influence in Brazil, Vargas secured funds to help build Volta Redonda in exchange for turning away from Germany and providing the Allies with airbases in northeastern Brazil. Vargas continued to welcome foreign investment to Brazil, but he wanted the investment on more favorable terms than in the past, and he made sure that the money generated more local economic growth.

Getúlio Vargas guided Brazil in new directions. He accelerated the process of modernization, but without breaking radically with the past. Known as the "father of the poor, and the mother of the rich," he played both sides of the aisle in Brazilian politics. By continuing to appease the powerful landowners and the military, and by offering reforms and access to the system (however limited) for the new urban groups, Vargas forged new political and social alignments. As had always been the case, the impoverished rural workers remained the ignored factor in national politics. And the system remained paternalistic and personalistic. Vargas now distributed the fruits of the system as had the coffee barons before him, and the monarchs before them. Furthermore, the centralizing and bureaucratic tendencies of the past accelerated after 1930 as Brazilian politics became even more statist. Stripping state governments of the power they had enjoyed under the First Republic, Getúlio appointed carefully selected "interventors" to run each state.

Never a true totalitarian, Vargas had been careful to allow limited dissent while maintaining selective repression and control. Isolating and neutralizing challenges became his forte. When many of the *tenentes* and their supporters rose up in a powerful rebellion in São Paulo in 1932, he managed to crush the revolt and check the power of the *tenentes* and *paulistas* who had opposed him. After a poorly planned Communist uprising in 1935, Vargas mobilized support for the regime by raising the specter of the radical left. By 1937, repression, prison, and exile had silenced the Communist Party and the left. The Integralists, Brazil's version of a fascist movement, made the same mistake as the Communists, attacking the national palace in a 1938 failed coup. Vargas now unleashed the repressive apparatus on fascists (who had cheered when the same measures had been used against the Communists), eliminating any further challenge from the radical right.

After juggling political interests in the early thirties under a republic, and as dictator after 1937, Vargas watched his regime begin to unravel in 1945. In the 1940s, pressure for elections began to build from all sides, and Vargas promised presidential elections for 1945. Adding to the pressure on Vargas were the events of World War II. In exchange for aid from the United States, Brazil sent 25,000 combat troops to fight in Italy in 1944 and 1945. As the only Latin American nation to contribute combat troops to the war effort, the nation and the military demonstrated Brazil's growing international presence. The Brazilian Expeditionary Force (FEB) also highlighted the contradiction between the war against dictators in Europe and the continued presence of dictatorship at home. How could Brazilians fight for the survival of democracy and against dictatorship in Europe without heightening the sense of their own lack of democracy? As in 1937, Vargas allowed political campaigning to begin for presidential elections.

As preparations took shape for elections, Vargas continued his juggling tactics, and many feared he would repeat the events of 1937, when he staged a coup shortly before promised elections. In 1937, the military had supported Vargas to crush perceived threats from both communists and fascists. In 1945, no such threats existed, and the military would not support any effort by Vargas to remain in power. Their desire for an industrialized and powerful Brazil had continued to grow, and many officers now saw the dictatorial regime as a stumbling block to development. In October, and with the blessing of many of the old political elite and the United States, the army forced Vargas to resign. Once again, the military exercised its self-prescribed role as the arbiter in national politics. Vargas quietly left for his ranch in southern Brazil, and the electoral campaign proceeded under a caretaker government.

THE AGE OF MASS POLITICS

The fall of the Estado Novo ushered in a new era of mass politics in Brazil, an "experiment in democracy" that would last two decades (a "Fourth" Republic from 1945-64). Although illiterates could not vote (and the illiteracy rate stood at about 50 percent), for the first time in Brazilian history the masses played a prominent role in the political system. Although a relatively small elite continued to control political power, all politicians, regardless of ideology, now had to campaign for the vote of the masses through modern political methods: rallies, radio, and newspapers. Under the First Republic, the percentage of the population voting in presidential elections reached 5 percent by 1930. By 1960, the figure had risen to nearly 20 percent. The experiment in democracy further reinforced the rise of populism and populist politicians that had begun in the 1930s. The economic nationalism and social welfare programs of populists had enormous appeal among the masses.

Another new feature on the political landscape was the formation of truly national political parties. Conservatives and Liberals in the nineteenth century had been loose coalitions of groups within the small political elite. Under the First Republic, some states (most notably São Paulo, Minas Gerais and Rio Grande do Sul) had statewide political machines, but no true national political parties existed. Four major parties emerged in the 1940s. The UDN (União Democrática Nacional) attracted the more traditional and conservative elements in national politics, while the PSD (Partido Social Democrático) appealed to more moderate and progressive elements in the traditional power structure. Building on the working-class reforms of Vargas, labor leaders and their political allies formed the PTB (Partido Trabalhista Brasileiro) to represent the interests of the Brazilian working class. While dominated by leftists, the PTB stood to the right of the fourth major party, the only party whose existence predated the Fourth Republic. Founded in 1922, the Brazilian Communist Party (PCB) had survived severe repression for more than two decades. Its leader, the former *tenente,* Luís Carlos Prestes, spent the last eight years of the Vargas regime in prison. The PCB and the PTB both competed for the support of the urban working class as their core constituency.

In the elections of 1945, the PSD candidate, Eurico Dutra, triumphed with 55 percent of the vote. After serving as Minister of War during most of the Estado Novo, Dutra was one of the most influential officers in the Brazilian military when he became a presidential candidate. (He officially retired from active duty just two days before his inauguration, although he wore his uniform for the ceremony.) In January 1946, he began a five-year presidential term. While finishing fourth in the election, the PCB garnered 10 percent of the vote and elected fourteen congressmen and one senator, making the PCB the most successful and the largest communist party outside the Soviet Union. With the onset of the Cold War, the Brazilian government (with U.S. support) banned the PCB in 1947 and persecuted its followers. With this important exception, the Dutra administration allowed a free press and open politics. A hesitant and cautious president, Dutra did not make any major changes in the political system. A new constitution went into effect in 1946 (Brazil's fourth since 1889), and the economic growth of the Vargas years accelerated.

Vargas, meanwhile, maintained a low profile and planned his return to power. During the late forties, Vargas cultivated the support of sectors of the traditional power structure while building an image as a man who had taken dictatorial steps to save the country from left-wing and right-wing extremists. Vargas the Democrat quietly convinced the military not to oppose him should he run for president in 1950. He then struck a deal with the populist governor of São Paulo, Adhemar de Barros. The latter had a powerful political machine in the most populous state in the nation, and in exchange for supporting Vargas in 1950, Getúlio promised to support Barros's presidential bid in 1955. With the support of the PTB and Barros, Vargas defeated the candidates of the PSD and UDN in 1950, gathering 49 percent of the vote. Five years after a military coup ended his dictatorship, Getúlio Vargas returned to the presidency with a resounding electoral victory.

The next four years would not be easy ones for the sixty-three-year-old Vargas, or for Brazil. Despite his electoral victory, opposition parties controlled the Senate and House, and they fought Vargas at every turn. Vargas turned to nationalism and urban workers as the pillars of his new politics. After a bitter campaign, the congress agreed to nationalize the petroleum industry, creating Petróleos Brasileiros (Petrobrás). Vargas continued to fight for greater Brazilian control over the economy through nationalization and state intervention in key sectors. Powerful business interests, multinational corporations, and foreign governments combined to fight back. Opponents of Vargas controlled almost all the major newspapers, magazines, and radio stations, and they attacked the president from all sides. Carlos Lacerda, a former communist turned radical rightwinger, became his most tenacious media nemesis. A bitter enemy of Vargas, Lacerda widely distributed radio and print commentaries that viciously attacked the president.

Perhaps the most difficult problem facing Vargas and all postwar governments was the economy, especially inflation. With a chronic annual inflation rate hovering around 20 percent, dissatisfaction ran high with regard to rising prices and lagging wages. Vargas could not satisfy the constant demands from all sectors for wage increases, and inflation crippled his administration. When Vargas suddenly raised the minimum wage for industrial workers by 100 percent in 1954, angry military officers, civil servants, and businessmen challenged the president. Corruption in the government also weakened Vargas and fueled the opposition as they unsuccessfully attempted to impeach the president.

By late 1954, the country had come to a political impasse (not unlike the Watergate scandal in 1974), with Vargas and his opposition in a deadlock. A dramatic attempt to assassinate Carlos Lacerda broke the deadlock. One of Lacerda's bodyguards, an air force officer, died in the attempt while the contoversial commentator suffered a slight leg wound. Investigations quickly tied Vargas' personal bodyguard to the attempt. (Vargas was probably unaware of the plot to kill Lacerda, and reportedly lamented, "that shot was aimed . . . at me.") Faced with this latest scandal, Vargas described himself as a president surrounded by a sea of mud. In late August the high command of the military gave him an ultimatum: resign or be forced out.

Facing the end of a long and brilliant political career, Vargas chose his most dramatic maneuver as his last. On the morning of August 24, 1954, he stepped into his bedroom in the presidential palace, placed a pistol to his heart, and pulled the trigger. Ironically, the shot that killed Getúlio Vargas also severely wounded his political opponents. A suicide note found next to Vargas eloquently and bitterly attacked his opponents and those who opposed Brazilian nationalism. The note ended dramatically, "I gave you my life. Now I offer my death. Nothing remains. Serenely I take the first step on the road to eternity as I leave life to enter history." As news of the suicide spread, a wave of sympathy for the fallen leader swept the nation as crowds attacked opposition newspapers, the offices of multinational corporations, and the U.S. embassy. Carlos Lacerda and his family took refuge on a military base. Once again, Vargas had turned the tables on his opponents and turned political defeat into victory.

The sudden death of the president touched off eighteen months of byzantine political maneuvering among the followers of Vargas (getulistas) and his enemies (anti-getulistas). Vice-president João Café Filho, a follower of Adhemar de Barros who had strong ties to the UDN, became acting president in a caretaker regime as the campaign for the 1955 elections took shape. A three-way split allowed the getulistas to capture the presidency and vice-presidency with just 36 percent of the total vote. The governor of Minas Gerais, Juscelino Kubitschek, and Vargas' controversial labor minister, João Goulart, led the successful PSD-PTB coalition ticket.

The great-grandson of a Bohemian immigrant, Kubitschek had been the mayor of the important industrial city of Belo Horizonte in the forties, before serving as the governor of the country's second most populous state in the early fifties. The dynamic JK (as he was called) forged a reputation as builder and industrial promoter. Although originally from a small town in the interior, and from a party (PSD) with a predominantly rural political base, Kubitschek represented the new politics of economic nationalism and modernization. A native of Rio

Grande do Sul, the young Goulart (nicknamed Jango) had been a close protegé of Vargas, serving as his minister of labor. With the death of his mentor, he stood to inherit the political mantle of the PTB, with its emphasis on social justice and working-class issues. The rural vote of the PSD combined with the urban vote of the PTB produced the Kubitschek-Goulart victory.

Facing another five years of a *getulista* administration enraged the defeated opposition. Led by the vitriolic media attacks of Carlos Lacerda, the anti-*getulistas* openly appealed to the military to stage a coup. A bizarre set of circumstances nearly gave Lacerda what he so desperately wanted. In November, Café Filho suffered a heart attack, and the president of the House of Deputies took over as acting president. An avowed opponent of the newly elected administration, Carlos Luz began to work with Lacerda and the opposition to stage a coup. High officials in the army, wishing to guarantee the constitutional process, quickly removed Luz from office, replacing him with the president of the Senate. In effect, the military staged a preventive coup, once again acting as the key power broker in the political process. In January 1956, the new administration took office.

Juscelino Kubitschek campaigned on the slogan "fifty years in five," promising to achieve fifty years of progress during his five-year term. Arguably, he succeeded. In many ways, Kubitschek is Brazil's greatest president. He was certainly the last freely elected president to serve a full term. While memories of Vargas and Goulart bring back bitter images to many Brazilians, and Dutra is remembered as a passive figure, Juscelino evokes nearly universal reverence for his extraordinary works. During the late fifties the Brazilian economy surged forward with the implantation of heavy industry (iron, steel, automobiles) and the expansion of basic infrastructure (roads, communications systems, and the construction industry). In a sense, the long, slow industrial growth that had begun in the late nineteenth century, and accelerated after 1930, took off during the Kubitschek years.

Juscelino's most vivid and enduring legacy rises on the plains of central Brazil. It was Juscelino Kubitschek who had the vision and the energy to construct a new national capital in the Brazilian interior. The idea of moving the capital into the interior dates back to the eighteenth century, but it was Kubitschek who convinced the legislature to accept the idea, and to fund it. Located near the boundary between Minas

Gerais and Goiás, 800 miles north of Rio de Janeiro, Brasília symbolizes
the emergence of a new Brazil. Between 1956 and 1960, Kubitschek
personally supervised the construction of this modern, futuristic city,
often flying from Rio to Brasília and back during the night to check on
the project. Inaugurated in April 1960, Brasília now has more than a
million inhabitants and has succeeded in drawing immigrants and
economic growth into the neglected interior as Kubitschek envisioned.
Brasília stands as Juscelino Kubitschek's most tangible political legacy.

DESCENT INTO CHAOS

Prohibited by the constitution from succeeding himself, Kubitschek
could not enter the 1960 presidential campaign. After decades of bitter
battles between the supporters and opponents of Vargas and his legacy,
a new figure, who seemed to promise an end to this struggle, appeared
on the national political scene. A political outsider, Jânio Quadros rose
to national prominence swiftly and unconventionally. Elected state
deputy in 1950 and then mayor of São Paulo in 1953, he won the
governorship the next year. Cultivating the image of an "anti-politician"
and an efficient administrator, Quadros had enormous appeal among
the middle and working classes. He vowed to sweep government clean
of the corrupt old boys, and he even brandished a broom as his symbol
during one campaign. With his charisma and voter appeal, Quadros had
parties coming to him offering their support and nomination. As one
historian has pointed out, Quadros had emerged as a legendary figure in
the late fifties, a symbol of the new, industrial Brazil, "a politician who
could direct effective government of a rapidly growing economy while
reconciling the social conflicts which it produced."

Quadros received nearly half the votes in the 1960 election, while
the traditional political parties of Adhemar de Barros and the PSD split
the other half. The UDN had nominated Quadros, but he made it clear
that he was in no way indebted to them. In January 1961, unencum-
bered by political debts or party ties, he took office. The separation of
the presidential and vice-presidential voting that had given Vargas a vice-
president from a different party now saddled Quadros with João
Goulart. As in the United States, the office of the vice-president has been
largely ceremonial and unimportant—under normal circumstances.

The abnormal circumstances of 1954 thrust the vice-president into a prominent role and highlighted the difficulties caused by a system that allowed for different parties to split the two offices. In 1961, extraordinary events once again demonstrated the flaws of this system.

In January 1961, Jânio Quadros raised hopes for the beginning of a new era in Brazilian politics. Within seven months, his bizarre behavior plunged the nation into a major political crisis that led to the collapse of the experiment with democracy. Quadros rarely held cabinet meetings, preferring instead to fire off notes on little slips of paper (*bilhetinhos*) to everyone, from cabinet officials to clerks, with his latest pet projects. He had a penchant for focusing on quirky issues. In a country famous for producing the world's smallest bathing suits, Quadros once ordered bikinis banned from Brazilian beauty contests. Two decades later, as mayor of São Paulo, he went on a campaign against motorists who parked on sidewalks, personally handing out the parking tickets. Throughout his long political career (stretching into the 1990s), Quadros was also rumored to have a major drinking problem.

Within a very short time, Quadros managed to anger just about every major political constituency. At odds with the legislature over some of his proposed programs, Quadros suddenly and unexpectedly resigned the presidency in August, just seven months into his term. No one, including Jânio Quadros, has ever offered a satisfactory explanation for the resignation. Although many believed he was bluffing and expected the resignation to be refused—returning him to office with broad support—his behavior and inaccessibility immediately after the resignation belie this explanation. In later interviews, Quadros enigmatically mentioned "personal demons" that plagued him. Whatever the reasons behind this strange behavior, it provoked a constitutional crisis.

The constitution called for Vice-President João Goulart to succeed Quadros, but powerful figures in the military high command quickly declared him unacceptable. Many Brazilians on the right and middle of the political spectrum saw Goulart as a communist sympathizer or communist at worst, and as too far to the left of center at best. Ironically, at the moment Quadros resigned, Goulart had just completed an official state visit to Communist China. The congress, and many political leaders, rejected the military's position and called for respect for the constitutional process. For nearly two weeks, the military and congress negotiated a solution to the impasse. A major split developed within the military between those who opposed Goulart and the "legalists" who believed that the constitutional process had to be respected. Unable to present a unified front, the military could shape but not control the outcome of the political crisis. While Goulart slowly made his way back to Brazil (via Paris, New York, Buenos Aires, and Montevideo), politicians and generals struck a compromise. Goulart would be sworn in, but his presidential powers would be curtailed through new legislation that created the post of prime minister. The compromise respected legal process (at least in theory) and made Goulart's inauguration more palatable to his enemies.

The military's hatred of Goulart must be seen in the context of their professionalization during the preceding decades. By the 1960s, the leadership of the Brazilian military had been forged out of the experience of combat in World War II, their alliance with the United States during the Cold War, and their education in recently created schools and courses for the officer corps. Especially in the Higher War College (Escola Superior de Guerra), the officer corps had developed and disseminated a vision loosely called National Security Doctrine. The military viewed Brazil as a frontline nation in the struggle between the superpowers, between an atheistic Communist East and a Christian capitalist West. Both were locked in a struggle to the death, as the East attempted to subvert the West through Communist infiltration and the destabilization of politics in the Third World. Fidel Castro and the Cuban Revolution after 1959 served as a powerful reminder to the anti-communist officers of the threat they faced. The left, furthermore, represented not only a threat to Brazil, but to the future of mankind. Goulart, with his support for leftist causes, shaped up as the symbol and leader of the forces of evil in Brazil. These officers feared another revolution in Brazil, one that could dwarf the bloodshed and conflict in Cuba.

During his two-and-one-half years in power, Goulart seemed to confirm all the worst fears of the military, the right, and ultimately, the center of the political spectrum. Many of the political problems stemmed from the economy's gradual disintegration. Kubitschek had left his successors with 25 percent annual inflation, a growing debt crisis, and balance-of-payments problems. In all fairness to Goulart, he inherited the bills for Kubitschek's grandiose projects. Under Goulart, however, the crisis intensified to intolerable levels. By early 1964,

Goulart had given up on efforts to impose an economic austerity program. Inflation approached 100 percent a year, foreign loans came to a halt, and the economy reached the collapse point. Goulart had no hope of cultivating political support from the right, and the economic crisis eventually destroyed any hope of building support among the middle class and the political center.

Goulart matched his mishandling of the economy with his political ineptitude. For the first fifteen months of his term, he focused his energies on regaining full presidential powers. In a plebiscite, the nation voted five to one in favor of restoring the old presidential system on January 6, 1963. Many Brazilians, it seemed, were willing to give João Goulart a chance to move the country forward. Unfortunately, the center did not hold as Goulart mismanaged the economy and moved to the left. As hopes for building a constituency in the political center seemed to diminish, Goulart eventually followed the advice of his most radical advisors and attempted to strengthen his support among the masses. Although numerically the majority, the urban and rural workers had never carried much political weight in Brazilian politics. In a serious political miscalculation, Goulart and his advisors believed that the rural and urban masses could be mobilized to check the opposition of the elites and the middle class.

In the early sixties, it did appear that a powerful leftist movement built on the support of the masses might revolutionize Brazil. In the Northeast, a young lawyer named Francisco Julião had begun organizing Peasant Leagues to challenge the powerful landowners. In the cities, labor had decades of experience organizing unions, and a number of union leaders were communists or leftists. Even the military counted among its ranks a number of senior officers who were marxists or leftists. In retrospect, we can see that nearly everyone overestimated the power and influence of the left in Brazilian politics. The left was weak, small, poorly organized, and badly divided over tactics. The PCB, sensing this, counseled Goulart to move slowly and with moderation.

In a series of desperate moves in the first months of 1964, Goulart staged mass rallies in several of Brazil's major cities. Attempting to mobilize mass support, the rallies featured speeches by radical politicians and the signing of nationalistic decrees that subverted the normal legislative process. While these rallies intensified political polarization, a series of conflicts with the military sealed Goulart's fate. Leftists had attempted to unionize the lower levels of the army and navy, and Goulart's interference into the military's efforts to discipline labor organizers and some mutineers tipped the delicate balance against him. In a final, desperate move to check the power of his enemies in the high command, Goulart made a televised speech to a group of sergeants, telling them that they should disobey their superiors if they believed their orders were not in the best interest of the nation. This call to insubordination set in motion a coup that toppled the regime.

MILITARY RULE

Conspirators in the military had been contemplating Goulart's overthrow for months. Initially, the plotters had planned a "defensive" coup to halt any attempt by the left to solve Goulart's problems through a revolutionary uprising. (The conspirators had maintained very close contacts with the U.S. embassy, and U.S. forces were poised offshore to come to the aid of the coup should they be needed.) Led by General Humberto Castello Branco, a veteran of the Brazilian Expeditionary Force (FEB) in Italy, the conspirators decided to take the offensive. As conditions deteriorated in March 1964, the key question became when to move into action. On March 31, after Goulart's speech to the sergeants, the commander of the army in Minas Gerais, General Olímpio Mourão Filho, took the initiative and ordered his troops to move on Rio. Joined by the garrisons in Rio, the movement gained force on the morning of April 1. When the armed forces in São Paulo joined the rebellion, they guaranteed the success of the coup. Goulart fled the country, never to return. While many Brazilians may have welcomed the downfall of the regime, few suspected that a new era in Brazilian history had begun (a "Fifth" Republic).

Brazil became the first of many Latin American democracies to fall under military rule in the late sixties and early seventies, and it would be one of the last countries to make the transition back to democratic politics in the eighties. The coup followed the long-established pattern of military intervention to resolve a political crisis. In 1889, 1930, 1937, 1945, 1954-55, and 1961, the military had acted as the key political broker, intervening in a moment of crisis and stepping back from power as the crisis passed. This time, however, the military remained in power for twenty-one years. Convinced of the ineptitude and irresponsibility of civilian politicians, and fearful of the rise of the armed left all across Latin America, the military high command decided to take charge of the country. In a sense, the military regime represented the culmination of the positivist ideal of the late nineteenth century: the creation of an authoritarian republic led by technocrats promoting industrialization. The military intervened with two primary objectives: to eradicate the left and to rebuild the collapsing economy. Military leaders split between political hardliners *(linha dura)* and moderates over how to achieve these goals.

Led by Castello Branco, who was named president, the moderates dominated the early years of the regime. Rather than shutting down civilian politics completely, the military attempted to "purge" the system of "undesirable" elements. Arrest and imprisonment of anyone perceived to be an opponent of the regime became common. Many fled the country. The military stripped thousands of civil servants, military personnel, and politicians of their right to participate in politics. Among those on the infamous lists of undesirables were Juscelino Kubitschek and Carlos Lacerda. Through selective repression, the military hoped to crush any threat of upheaval and to cow the legislature and judiciary into submission. On the economic front, the new regime tried to institute an orthodox economic program to bring down inflation and restore Brazil's foreign credit. Despite some initial success, the regime by 1967 found liberal economics wanting and embarked on a program of massive state intervention in the economy.

By 1968, growing political opposition, even from the traditional elites, increasingly called for the military to return power to civilians. In the face of demonstrations, a recalcitrant legislature, and the rise of a small guerrilla movement, the hardliners emerged with the upper hand in the military high command. In 1967, General Artur Costa e Silva succeeded Castello Branco (who died in a plane crash a few months later). Costa e Silva came from the camp of the hardliners. A *tenente* in the 1920s, he had been Castello Branco's minister of war.

A series of events in late 1968 brought the hardliners' wrath down on the opposition and triggered a coup within the regime. Massive student protests disrupted and challenged governments across the western world in 1968, and university students in Brazil also mounted huge demonstrations against the generals. The military responded with force. In August and September, Costa e Silva suffered a series of strokes. With the president incapacitated, the three military cabinet ministers (army, navy, air force) took charge, ignoring the constitutional line of succession. In the midst of the succession crisis, urban guerrillas kidnapped U.S. Ambassador Charles Burke Elbrick and demanded a ransom and the release of political prisoners. Meanwhile, the efforts of the regime to imprison a congressman who had openly criticized the military provoked a confrontation with the legislature.

The generals saw chaos and communists all around them and they cracked down, initiating intense repression to crush the opposition. In December 1968, they shut down congress. The military triumvirate issued a new constitution that concentrated power in the executive and named a new president, General Emílio Médici, the former head of the intelligence agency. Between 1968 and 1973, Médici and the hardliners unleashed the systematic and widespread use of torture and repression to silence their opponents. Thousands suffered at the hands of the torturers, and hundreds died at the hands of the military. These were the darkest and most sinister years in Brazilian history.

The use of torture was not new in Brazil. For decades, if not centuries, the police have routinely beaten suspects. The use of torture in the 1960s and 1970s differed from the methods of the past in three important ways. First, under the military its use became more systematic, sophisticated, and widespread. An assortment of security forces—civil, military, and political—cast their net widely, and often. Second, the focus of the torturers was not the common criminal suspect, but those suspected of "political" crimes, a term notable for its ambiguity and flexibility. Finally, the principal targets of the repressive apparatus were not the lower classes, but professionals, the middle class, the educated. In short, those with the education, the time, and the financial resources to organize, and oppose the regime.

The guerrillas presented the most difficult challenge to the military, yet within a few years they had been entirely wiped out. Sophisticated technology in the hands of a very powerful central state, the advice and close assistance of the United States military, torture, and the ineptitude of the guerrillas all contributed to the insurgency's demise. Urban workers and students also stood up to challenge the generals and paid dearly. The government eventually shut down the national student union and took direct control of labor unions. Universities purged their faculties of the ideologically "suspect" and silenced anyone who criticized the regime. Large numbers of prominent Brazilian academics and artists went into exile in other Latin American countries, the United States, and Europe.

Spectacular economic growth accompanied the success in suppressing political opposition. The years of repression coincided with the years of the so-called Brazilian miracle, during which the economy grew at an annual rate of 11 percent, faster than any country in the world. Ironically, this phenomenal growth came under a right-wing military regime and through widespread state intervention in the economy. The staunchly nationalistic military wanted to make Brazil a world power and understood that a strong industrial economy held the key to their goal. Although it welcomed foreign investment and attracted billions of dollars, the regime channeled that investment into sectors of the economy considered critical for development. Among other things, the combination of massive state spending and foreign investment built the Trans-Amazon Highway, the world's largest hydroelectric dam (Itaipú), and a nuclear power program.

By 1973, the military seemed to have exceeded even their own expectations. They appeared to have control over the political system—suppressing any dissent—and the economy was expanding at an extraordinary pace. The technocratic authoritarian state seemed to work. The sentiments of the generals were summed up in a slogan the government heavily promoted at the time, "No one will hold this country back." In retrospect, the inauguration of a new president and the first oil crisis in 1973-74 marked an important turning point for the political system and the economy.

Moderate forces within the military brought General Ernesto Geisel to the presidency, and they began a gradual process of political opening that took more than a decade to complete. The son of German immigrants, Geisel initiated a series of reforms that gradually allowed limited political organization and elections. This effort at political "decompression" or *abertura* (opening) had its ups and downs. When the "official" opposition party, the Movimento Democrático Brasileiro (Brazilian Democratic Movement, or MDB) won major electoral victories in the mid-seventies, the government issued a series of laws in April 1977. Known as the April Package, they imposed severe restraints on the regime's opponents. A dilemma faced the military moderates as they sought to open the system. Each time they allowed any elections, the opposition made impressive gains. These elections made it clear that Brazilians, even those who had initially welcomed the coup, wanted the military to return power to civilians.

Growing economic problems complicated *abertura*. The enormous industrial and economic expansion of in the late sixties and early seventies made the country heavily dependent on petroleum and its byproducts. When the first oil crisis erupted in October 1973, Brazil was importing 80 percent of its oil. Although by the 1980s the military regime reduced this figure to 50 percent by increasing domestic production of petroleum, the high price of oil seriously crippled the economy. The Brazilian "miracle" had also been built on heavy international borrowing, and in the aftermath of both the 1973 and 1979 oil shocks the regime tried to borrow its way out of the crises. By the early 1980s, the $3 billion foreign debt of 1964 had escalated to more than $100 billion. The steep rise in interest rates in the late seventies fueled the rapid rise in debt and brought the country to the brink of default. Although the economy continued to grow at healthy rates until 1982, the burden of debt payments and triple-digit inflation became the most visible evidence of a growing economic crisis.

In the late seventies, growing economic problems and conflict over the direction of the political opening seriously divided the military. Geisel managed to handpick his successor, but only after a major revolt within the regime. General João Baptista Figueiredo proved to be a poor choice to finish the delicate process of *abertura*. Figueiredo's presidency began with promising developments. He declared a general amnesty for all political crimes since 1964 and allowed exiles to return home. The government released the last few political prisoners, and official censors finally left the pressrooms and television studios. The Figueiredo government issued guidelines for the formation of new political parties and for open election of governors in 1982.

The military game plan for *abertura*, however, came unraveled in the early 1980s, in the face of massive public demonstrations and economic crisis. Inflation surged forward in 1980 at close to 100 percent and continued its upward trend, reaching levels far higher than those of the 1963-64 crisis. In 1982, Brazil halted all payments on the principal of its huge foreign debt, and the economy entered into a severe recession. The country could barely make interest payments on the debt, and the billions of dollars in debt service payments drained the economy of precious resources.

With the threat of leftist revolution long since crushed and with a battered economy, the military regime found itself severely discredited in the eyes of most Brazilians. In 1984, millions of Brazilians took to the streets and demanded immediate direct elections (*diretas já*) for president. Although the government managed to fend off the calls for direct elections, the massive public movement helped split the government party. A large share of the government's electors in the electoral college defected and voted with the opposition in 1984, defeating the "official" civilian candidate for president. Previously, under the close control of the regime, the college elected Tancredo Neves to become Brazil's first civilian president since 1964. To share in the spoils of political patronage, the block of politicians who defected from the government camp struck a deal with Neves and the opposition coalition. In what seemed at the time to be a largely ceremonial gesture, the coalition made José Sarney Neves's vice-presidential running mate. Sarney, a long-time leader of the government party in the Senate, had played a key role in leading the renegade politicians into the opposition coalition.

A "NEW" REPUBLIC

A native of the politically powerful Minas Gerais, Tancredo Neves seemed the perfect leader for the transition back to fully democratic politics. He had begun his national political career as a cabinet minister under Vargas in the early 1950s, had served as governor of Minas Gerais and as prime minister during the Goulart administration in 1961. For two decades he had been a prominent member of the MDB who opposed the military government, yet he was universally recognized as a "responsible" (that is, not radical) politician by the generals. He was the quintessential insider opposition figure. Tragically, he was also desperately ill. On March 14, 1985, the evening before his scheduled inauguration, Neves entered the hospital with a perforated intestine. For more than a month, he lingered near death. Eight operations could not save him from the infection that attacked his seventy-four-year-old body. Neves died in late April and José Sarney was sworn in as president.

While Tancredo Neves had spent years carefully cultivating and constructing a broad political base on his way to the presidency, José Sarney unexpectedly backed into the presidency after cutting ties with most of his former political allies. The sudden turn of events denied the opposition movement control of the presidency at the very moment of their triumph, and it brought to power a prominent politician who had long represented the military regime. Sarney had angered the regime by defecting to the opposition, among whom he had no political base and faced hostility and suspicion. Furthermore, he inherited a cabinet and government officials already appointed by Neves. Rather than a consensus candidate with broad political support, Brazil now had a president with enemies on all sides and virtually no political base.

Sarney immediately faced two momentous problems: the economic crisis and the need to continue the transition to a fully democratic regime. Inflation in 1985 approached 300 percent, the currency lost value on a daily basis, the foreign debt continued to mount, and strikes broke out across the country as workers demanded higher wages. In a drastic effort to stabilize the economy, Sarney introduced the Cruzado Plan in February 1986. The plan froze prices and wages and converted the old currency (the *cruzeiro*) into *cruzados,* at a rate of 1,000 to 1. The shock effects of the plan brought Sarney to the peak of his popularity. Inflation ground to a standstill for a few months. Unfortunately, the government kept the lid on the freeze too long, waiting until after the November 1986 elections to ease it. Tremendous consumer demand spurred imports, causing a balance-of-trade problem. As costs soon outpaced prices, producers held back goods, causing artificial scarcities. When the government did unfreeze prices and wages at the end of 1986, inflation exploded again.

The foreign debt also posed challenges for the Sarney government. Interest payments alone gobbled up nearly all of the country's huge trade surplus, draining the economy of badly needed capital. The government incurred large deficits in public spending, and foreign banks refused to extend new loans until the government implemented an economic austerity program. Austerity required a strong president with a mandate, and it would cause serious political and social unrest. Unwilling to risk facing the domestic costs of austerity, Sarney announced a moratorium on debt payments in February 1987. By early 1988, the government suspended the moratorium and resumed negotiations with the banks.

Brazil badly needed new loans and investment, and these would only come after it renegotiated the foreign debt and implemented an austerity program at home.

A constitutional convention further restricted Sarney's ability to maneuver. The congress elected in November 1986 served also as a constituent assembly and, throughout 1987 and 1988, it slowly drafted a new constitution (for a "Sixth" Republic). Facing a hostile assembly, Sarney relied heavily on support from the military (and the spoils of political patronage) to guarantee himself a five-year term. The military feared open national elections for president and wanted to delay direct presidential elections for as long as possible. Unfortunately, José Sarney spent his first three years maneuvering to stay in power for a full five years. Rather than the expected leadership and drive to put the country back on a prosperous, democratic track with Tancredo Neves, Brazil drifted for five years under the uninspired leadership of José Sarney.

The election of Fernando Collor de Mello in late 1989, and his inauguration in March 1990, marked the completion of the long and difficult process of *abertura*. Finally, Brazilians had the opportunity to elect their president through the ballot box rather than through imposition by a small clique of generals. More than 80 million Brazilians voted in the presidential election, the vast majority for the first time. In the first round of voting, Brazilians shunned the faces from the past—many of whose political careers dated back to the pre-1964 era—and narrowed the field to the forty-year-old Collor and the forty-four-year-old Luís Inácio da Silva (better known as Lula). The runoff in December 1989 became a clear choice between the leftist, interventionist program of Lula, and the anti-interventionist, free-enterprise message of Collor.

In his first two years in office, Collor initiated some of the most sweeping and far-reaching changes in modern Brazilian history. He implemented an economic shock plan that brought inflation down but failed to contain it. More important, he began a process aimed at drastically removing the state's role in the Brazilian economy. The government has begun to sell off state-controlled enterprises, reversing fifty years of growing economic intervention, and to dismantle the protectionist trade policies that have been the staple of both left-wing and right-wing nationalists. A fundamental shift in the nature of political debate may well be Collor's most enduring legacy. By taking the offensive, Collor focused the debate on *how* to transform Brazil, rather than whether it *should* be transformed.

The great hopes that millions of Brazilians had for the Collor presidency soon disappeared as the economic shock treatment failed to halt four-digit inflation rates, and as corruption scandals shattered the government. In mid-1992, the president's brother, Pedro, made startling accusations of drug use by Fernando and revealed a massive kickback scheme coordinated by the president's close friend and campaign treasurer, P. C. Farias. Over the next six months, legislative investigations uncovered an influence-peddling scheme that involved hundreds of millions of dollars. Farias apparently extorted money from large corporations in exchange for political favors. Through "phantom" bank accounts, he channelled millions to Fernando Collor for his personal use. Collor's wife, Rosane, had a personal clothing expense account in the vicinity of $20,000 a month, and the couple spent millions landscaping their backyard with waterfalls and exotic plants. Even in a system that has long been characterized by politicians' profiting from their public office, the scale of the corruption shocked and angered Brazilians.

In December 1992, the congress impeached Collor and swore in his vice-president, Itamar Franco, to serve out the two remaining years of Collor's term. The impeachment offered Brazilians a rare chance for a detailed glimpse at the mechanisms of corruption and patronage in their political system, mechanisms Collor exploited to degree exceeding anything previously seen in Brazil. Despite the damage Collor had done to the nation and its political system, his impeachment (the first of a president in the history of the Americas) did demonstrate that a major political crisis could be dealt with through constitutional processes. The politicians handled the crisis, and the military maintained a notably low profile.

President Franco's greatest achievement may have been paving the way for the election of his successor, Fernando Henrique Cardoso. One of Latin America's most prominent intellectual figures, Cardoso was trained as a political sociologist at the University of São Paulo in the late fifties and early sixties. Once a member of the Communist Party (PCB), Cardoso spent part of the sixties and seventies in exile, coauthoring the most influential book on Latin American development in the past thirty years, *Dependency and Development* (1967). During the late seventies, he entered politics, eventually becoming a senator from the state of São Paulo and an unsuccessful mayoral candidate for the city.

In early 1994, in yet another effort to combat runaway inflation and the debt crisis, Franco chose Cardoso as his finance minister. Cardoso and a very fine team of advisors put together the Real Plan. On July 1, 1994, this plan created yet another new currency, the *real* (plural, *reais*) and put into place a series of measures to reduce inflation without wage or price freezes. Inflation dropped from 45-50 percent per month in early 1994 to around 1-2 percent per month over the next two years, giving Brazilians their lowest inflation rates in decades. The success of the plan made Cardoso a national hero and the leading contender for the presidency.

When the electoral campaign began to take shape in early 1994, Lula lead all polls with some 40-45 percentage points. Cardoso was the preference of only about 20 percent of those polled. By September, those numbers were reversed. Cardoso forged a coalition of his Brazilian Social Democratic Party (PSDB), the conservative Liberal Front Party (PFL), and several other parties. The marxist intellectual convinced business-people and conservatives that his views had evolved and were close enough to theirs to gain their support. They certainly preferred Cardoso to the more radical Lula. With nearly 55 percent of the total vote in the November elections, Cardoso scored the most impressive electoral victory in a presidential election in forty years.

Inaugurated on January 1, 1995, President Cardoso has the political and intellectual skills to make his mark as one of the great presidents in Brazilian history. During his first year in office, he forged an impressive congressional majority coalition that passed fundamental reforms to the constitution. Four presidential administrations (one military and three civilian) have failed to conquer the major problems that Brazil has faced throughout the 1980s and 1990s: the debt crisis, political drift, and a growing social crisis. Should Cardoso continue the consensus-building, he could lead Brazil into a new era, in which the nation could finally address these fundamental challenges and begin to overcome them. (In early 1997, congress passed an amendment to the constitution allowing for the re-election of the president for a second term, opening the way for Cardoso to serve eight years.) The last years of the century may be the most democratic period in Brazilian history and an era of renewed economic growth. The country appears to be back on the road to the future.

CONTINUITY AND CHANGE

At the end of the twentieth century, the major challenge facing Brazilians is how to confront the enduring legacies of the past, and how to transform them. Brazil has changed dramatically during the past two centuries, yet many of the problems facing late-twentieth-century Brazil have roots deep in the nation's past. Although Brazil today has all the trappings of modern, mass democratic politics, elite groups continue to play the decisive role in the political system. A central state that has slowly grown and asserted control over an immense nation shows few signs of relinquishing or redistributing power. The military, despite its return to the barracks, shows no signs of renouncing its historical role as the ultimate power broker in the political process. The public participates in the political process, but it continues to play a role that is more reactive than active. The appalling wealth inequities reinforce and perpetuate the maldistribution of political power. A very small (and overwhelming light-skinned) portion of the population continues to control most of the wealth and power.

In the last decade of the twentieth century, Brazilians continue to grapple with the burdens of the past as they head toward the twentieth-first century. Gripped by self-doubt that is forged out of the cyclical ups and downs of the nation, some Brazilians look optimistically to the future. Others, at times, are overwhelmed by the burdens of the past. In an earlier moment of pessimism, the renowned writer, Mário de Andrade, once bitterly complained that, "My past is not my friend. I distrust my past." The burdens of this past grew out of the unique collision of three worlds that gave birth to Brazil. The struggle between the Portuguese masters and the Native American and African populations they subjugated created Brazil. Although the Portuguese had more power and clearly played the dominant role in the struggle, Native Americans and especially Africans played crucial roles in the creation and construction of a fascinating cultural mosaic that is neither European, American, nor African. The enormous imbalance in power among the groups, however, produced a society scarred by profound divisions and inequities that have persisted in various guises for centuries.

Despite centuries of far-reaching changes in Brazil, the divisions and inequities forged during three hundred years of colonialism continue to plague Brazilians. Gone is the Portuguese colony that produced cash crops bound for Europe, a colony in which a small group of landowners dominated a rural population and a slave majority. The old Brazil has been transformed by urbanization, industrialization, the rise of new social groups, and new political ideologies. Yet the modern, urban, industrial, democratic Brazil of the late twentieth century continues to grapple with this legacy of profound inequities.

Over the last five centuries, Brazil has been constructed by a diverse mixture of peoples: Europeans, Amerindians, Africans, rich and poor, peasants and workers, landowners and industrialists. They created a single nation composed of astonishingly diverse regions, a racially mixed yet bitterly divided society, a democratic political system dominated by elite groups, and an enormously productive industrial economy surrounded by enormous poverty. Brazilians created a nation of paradoxes. The remainder of this book surveys this intriguing and paradoxical country against the backdrop of the constant tension between past and present, between the country Brazil once was and the country it might become.

The Brazilian Archipelago

> The astonishing sociological phenomenon of
> Brazilian unity is nourished by this very diversity
> of regions—Brazils in the plural—which inter-
> mingle and are fulfilled in one single Brazil.
> — Gilberto Freyre

LIKE THE ISLANDS OF SOME HUGE ARCHIPELAGO, strikingly distinct
regions spread across a vast sea of geographical diversity to form the
Brazilian nation. Across the centuries, regionalism has run parallel and
counter to the central government's efforts to impose a common
language, religion, and culture. On one level, the process of homogeni-
zation has triumphed and created a nation-state. On closer inspection,
one sees that intense regional identification and loyalties have survived
and even thrived. Much like other nations of continental dimensions—
Russia, Canada, the United States, China, and India—Brazil has
struggled to impose a national unity on regional diversity. Brazilians
have perhaps been more successful at forging a stable and subtle balance
between a national culture and identity, and distinct regional cultures
and identities, than any other large nation.

Brazilians are acutely aware of regional differences, and they
identify closely with their regional origins. Not simply a matter of
geography and culture, regionalism in Brazil has become a state of mind.
Just as North Americans are conscious of the accents, ethnicity, and
cultural traits of Southerners, Westerners, New Yorkers, and the like,
Brazilians peg each other as northeasterners, southerners, and natives of
Rio de Janeiro or São Paulo. Even more so than Americans, Brazilians
define themselves by their regional roots.

The emergence of sharp regional distinctions should not be surprising in countries of continental dimensions, and Brazil truly has continental dimensions. The fifth largest country in the world (at 3,286,426 square miles, or 8,506,663 square kilometers), Brazil is slightly larger than the forty-eight contiguous United States. Covering nearly half the landmass of South America, and a third of all Latin America, Brazil is three times the size of Argentina, and four times the size of Mexico. All of Europe, from Scandinavia to the Mediterranean, and from the Atlantic to the Urals, could fit inside Brazil with room to spare. With a quarter of the world's proven iron ore reserves, huge deposits of manganese, uranium, bauxite, diamonds, and gold, Brazil has many of the key strategic resources to supply a modern industrial economy. Its great weakness is the lack of large petroleum and coal reserves. (The small coal deposits are of poor quality, and the country imports 50 percent of its petroleum.)

Unlike other nations of continental dimensions, Brazil has only recently begun to open up its interior. For centuries, European colonizers and their descendants settled along the 4,600-mile coastline and did not venture very far into the interior. They left the vast interior to the highly dispersed Amerindian peoples. As recently as 1980, nine of every ten Brazilians lived in the coastal Atlantic states (and in the state of Minas Gerais, whose eastern boundaries lie some one hundred miles from the Atlantic). The narrow coastal plain runs up against the Guiana Highland north of the Amazon, and against the Brazilian Highland south of the Amazon. From southern Bahia to Porto Alegre, this narrow coastal belt squeezes between the Atlantic and a steep escarpment with an average elevation of 2,600 feet. An enormous and uneven plateau (covering more than 60 percent of the nation's territory) slopes gradually away from the coastal escarpment into the interior all the way to the Andes in the west, to the prairies in the south, and to the Amazonian rain forest in the north.

Three major river systems cut across the Brazilian interior. In the west, the tributaries that combine to form the Plata system begin very near the tributaries of the Amazon. They merge with the tributaries of the Paraná River in the South and empty into the south Atlantic as the La Plata River. The 1,800-mile-long São Francisco River originates on the southern fringe of the Highland and then runs northward parallel to the coast on the western side of the escarpment and emptying into the Atlantic on the northeastern coast. In the North, the massive "river-sea" (as the early explorers called it) known as the Amazon contains nearly 25,000 miles of waterways that span more than one-third of the nation. Rainforest covers most of the Amazon basin, and it covered most of the coastal plain before the Europeans came. The vast interior highland consists largely of hot, dry savanna and scrub forest. In the South, hilly uplands and rolling prairie stretch across the states of Paraná, Santa Catarina, and Rio Grande do Sul.

Brazilian regionalism has its roots in geography and in patterns of colonization. In the sixteenth and seventeenth centuries, the Portuguese constructed their colonial plantation society on the rich coastal plain in the *Northeast*, centered around what are now the states of Pernambuco and Bahia. With the discovery of gold and diamonds in the eighteenth century, the axis of power and the focus of colonial development shifted to the *Southeast*, a region that today includes the states of Espirito Santo, Minas Gerais, Rio de Janeiro, and São Paulo. Effective settlement and development of the uplands and prairies of the *South* (Rio Grande do Sul, Santa Catarina, and Paraná) did not really take shape until a new wave of European colonization in the last half of the nineteenth century took place. The Amazon basin in the *North* was sparsely settled until very recent times, and the *Center-West* has just begun to boom with Brazil's "westward movement." In many ways, Brazil looks very much like the United States in the early nineteenth century: just beginning to embark on westward expansion and on the effective settlement and exploitation of a vast interior.

A combination of geographical, economic, ethnic, linguistic, and cultural traits characterizes each of these five regions (Northeast, Southeast, South, North, and Center-West). Much like that of Spanish America, Brazilian regionalism has deep roots. Unlike Spanish America and Anglo America, Brazil did not fragment into multiple nations after the wars for independence in the nineteenth century. Brazil has been able to reconcile pronounced regional diversity with a very strong sense of national unity and identity. While there is one Brazil, there are also many Brazils.

THE NORTHEAST

The Portuguese implanted their first major colonial settlement on the northeastern coast of Brazil, and, from the late sixteenth to the late seventeenth century, sugar and slavery made the Northeast the richest colony of any European overseas empire. Today, the Northeast is the poorest and most backward region of Brazil. A comparison of the Northeast with the Southeast becomes something of a southern hemispheric version of the relationship between North and South (which are, in more precise geographical terms, Northeast and Southeast) in the United States. (Everything below the equator, after all, is reversed.) While the Southeast has been the economic heartland of Brazil since the eighteenth century, the Northeast has suffered from economic underdevelopment, widespread poverty, and low social indicators. The inequalities, in fact, are even greater than those between the U.S. North and South, as the Brazilian Northeast has fallen even further behind the Southeast in the last century.

With some 18 percent of Brazil's landmass and 29 percent (42 million) of its population, the Northeast encompasses two civilizations, two contrasting personalities. Fertile plains, sugar plantations, large urban centers, and African culture have shaped the civilization of the coastal belt. Barren soils, cattle ranching, drought, and the native culture have molded the civilization of the backlands. Sugar and slavery built the coastal civilization. Livestock and the *caboclo* forged the civilization of leather in the interior. Facing the Atlantic and West Africa, all of the states in the region are coastal, stretching from Maranhão in the north; through Piauí, Ceará, Rio Grande do Norte, Paraíba, Pernambuco, Alagoas, and Sergipe; to Bahia in the south. Yet each has its harsh and forbidding backlands.

While Maranhão and Piauí merge into the Amazon basin on their northern and western boundaries, three major geographical zones cut across all nine states. The rich coastal belt has long been the area of densest settlement, especially in the old sugar plantation crescent, which stretches from Pernambuco to Bahia. Originally covered by dense forest (*mata*), the rich coastal plain was cleared by the Portuguese, who planted sugar in vast quantities beginning in the late sixteenth century. In the fertile hinterland around Recife in Pernambuco, and around Salvador in Bahia, the Portuguese created the first great plantation society in the New World using African slave labor. Despite the relative decline of the Brazilian sugar economy since the late seventeenth century, sugar remains the major crop of the Northeast, and Brazil continues to be the world's largest exporter.

Further inland comes the *agreste*, with somewhat higher elevations, less regular rainfall, and lots of small subsistence farms. The *agreste* forms a transitional zone between the rich coastal plain and the backlands (or *sertão*) to the west. Semi-arid, and plagued by long droughts, high temperatures, and poor soils, the *sertão* forms a harsh and forbidding environment. If the Northeast is the rough equivalent of the U.S. South, then the *sertão* is Brazil's Appalachia—with even more severe problems. Extended droughts have occurred every eight to fifteen years since the eighteenth century in the so-called polygon of drought. According to some accounts, as many as half a million (50 percent) of Ceará's inhabitants perished in the 1877-79 drought. Long periods without rainfall are punctuated with sudden downpours that turn the barren backlands into lush, green fields and forests for brief periods. In this "green hell," the tough *sertanejos* (backlanders) raise livestock, cotton, and sisal.

The legacy of Portuguese colonialism weighs heavier on the Northeast than any other region of Brazil. From the mid-sixteenth century, when the Portuguese established their colonial capital at Salvador, until the late seventeenth century, sugar planters imported hundreds of thousands of Africans to labor on the expanding plantations. This colonial system left a deeply entrenched legacy of large landed estates, which were concentrated in the hands of a small white elite that controlled the single-crop economy, regional politics, and society. After the Dutch invasion and occupation (1624-54), this monocultural economy went into a long decline in the face of competition from new plantations in the Dutch, French, and British Caribbean. With the gold rush in the Southeast after 1700, the Northeast continued to lose importance, a phenomenon reinforced by the transfer of the colonial capital to Rio de Janeiro in 1763. The region's day had passed, and despite some efforts to recover, elements of the colonial legacy have persisted to the present.

The colonial sugar and slave economy created the most distinctive features of the northeastern coast and stamped the African influence indelibly on the region. Three centuries of the slave trade made the coastal zone the most Africanized section of Brazil in both racial and cultural terms. The latest Brazilian census (1991) classified 70 percent of the population as black or mulatto, compared with 33 percent in the Southeast and 17 percent in the South. In Salvador, the center of the old plantation society, virtually everyone has some African ancestry. As one moves away from the coastal zone, the population's racial heritage is derived more from a mixture of Europeans and Indians than Europeans and Africans.

Recife and Salvador became the major northeastern ports and the urban centers for the sugar civilization. Today, exquisite baroque mansions and cathedrals bear witness to the wealth and power of the *senhores de engenho* (sugar planters) who made Salvador and Recife their urban base of operations. Built at the confluence of several rivers on the Atlantic coast, Recife has long styled itself as the "Venice" of South America. Although initially the commercial center for the neighboring Olinda, Recife long ago eclipsed the picturesque Olinda not only as the major commercial center of Pernambuco, but also as the capital. With a population of more than one million today, Recife remains one of the two most important urban centers of the Northeast.

If it is true that Brazil's soul is African, then that soul resides in Salvador. Brazil's colonial capital from 1549 to 1763, and for centuries its major port of entry for slaves, the city of Salvador—more than any other Brazilian city—blends European and African traditions. While the exterior of the dozens of baroque churches—their straight lines, sober pastel façades, and square towers—remind one of Portugal, the culture and personality of *baianos* (those from Bahia) hearken back to Africa.

The diet, dress, music, and religion of the Northeast have been shaped by centuries of African influence. Much of the basic diet of the region (black beans, greens, okra, pork, rice, fried foods) blend African and American cooking and foods. Traditional musical instruments and many popular rhythms have deep African roots. Black women in white turbans and flowing dresses provide one of the classic images of Salvador. The sight of these *baianas* selling sweets and frying foods in dendê oil on the streets of Salvador is like scenes from the streets of Lagos or Luanda.

Religion, however, provides the most striking and visible reminder of the Northeast's African heritage. Over centuries, the spiritism of African slaves and their descendants slowly blended with Roman Catholicism to produce a fascinating and powerful religious mixture. As one astute French observer has noted, Catholicism in the Northeast is a white mask over a black face, a fine overlay on a deeply entrenched spiritism. Slaves and their descendants identified their many African deities with the saints and the saintly in the Catholic Church. In the

Virgin Mary, for example, they saw Iemanjá, the goddess of the sea. Xangô, the god of lightning, became St. Jerome; and they equated Oxalá, the god of the sky and procreation, with Jesus. Despite centuries of efforts by secular and religious authorities to suppress African spiritism, the practices survived and spread, even beyond the slaves and free blacks. Known by various names (*macumba* in Rio de Janeiro, *candomblé* in the Northeast), spiritism has reached a close accommodation with Catholicism and permeates virtually all social classes in the Northeast and Southeast, especially around Salvador and Rio de Janeiro.

Extreme and widespread poverty are the worst legacies of colonialism. Centuries of slavery, concentrated land holdings, and economic decline have left northeastern Brazil with the worst living standard of any area of Latin America. The region produces 15 percent of Brazil's Gross Domestic Product, yet it has nearly 30 percent of its population. At just 40 percent of the national average, income per capita in the Northeast is the lowest in Brazil. The per-capita income of the Center-West, the next poorest region of the country, is double that of the Northeast. Fully half of all poor Brazilians are *nordestinos*. In a country with a literacy rate of 80 percent, the rate for the Northeast stands at 60 percent. Around 88 of every 1,000 infants born each year in the Northeast die before reaching their first birthday— nearly double the infant mortality rate for the nation (50 per 1,000). (The comparable rate for Cuba—the lowest in Latin America—is 11 per 1,000 in 1989, virtually the same as the United States.) The regional inequities are even more pronounced when one compares the Northeast with the South and Southeast.

In a region that continues to be predominantly rural (60 percent of the population works in agriculture), more than two million peasants have no land, and the concentration of land in fewer and fewer hands has intensified in the past few decades. A mere 3.5 percent of all agricultural properties cover more than 60 percent of all arable lands. (That is, a small number of very large estates take up the majority of the agricultural land in use.) As a result, rural social and economic structures have changed very little for centuries. Large estates (*latifundios*) dominate the countryside, and the poor peasants who work on these estates have traditionally staked out tiny plots (*minifundios*) on the fringes of the plantations where they live and produce their subsistence crops. Powerful landowners in the interior continue to control local politics and society, and millions live in abject poverty in the countryside. The landholding system has long been geared toward the production of cash crops, creating a bitter irony: This rich agricultural region has to import 70 percent of its foodstuffs from other regions of Brazil.

For centuries, backward and oppressive conditions in the Northeast's interior have produced sporadic outbursts of violent, and unsuccessful, social protest. In the nineteenth and early twentieth centuries, in particular, bandits in the mold of Robin Hood prowled the backlands. The most famous of these folk heroes, Lampião (the Lamp, the nom de guerre of Virgolino Ferreira da Silva), led his band through the interior in the 1920s and 1930s and built a legend much like that of Bonnie and Clyde in the United States. Killed by police in an ambush in 1938, he lives on in popular music and folk literature. Perhaps the most powerful and famous of the explosions of social protest occurred at the small town of Canudos in the backlands of Bahia in the late 1890s. As we saw in chapter 1, a religious mystic gathered thousands of the downtrodden around him and eventually challenged the local government's power. Only after a seven-month siege did the Brazilian army crush the movement and restore order. Lampião and Canudos represent but two of the many sporadic outbursts of protest against oppressive conditions in the interior.

Much like the American South after World War I, the Brazilian Northeast has long been exporting its human capital to other parts of Brazil, particularly the industrial Southeast. While thousands left the American South for the cities and factories of the North, Midwest, and West in the first half of this century, the flow has now reversed as the region imports human capital. The Brazilian Northeast continues to export its people in staggering quantities. In the 1960s, some 18 million *nordestinos* abandoned their home, followed by another 24 million in the 1970s. They migrated to all regions of the nation, some seeking land in the Amazon basin and the Far West, others seeking jobs in the factories of the Southeast. Many helped build Brasília in the late fifties and early sixties.

The flood of migrants into São Paulo and Rio de Janeiro has helped foster the stereotype of the dark-skinned country bumpkin from the Northeast with a sing-song, drawling accent. Recent sociological surveys have shown that discrimination and prejudice against the *nordestino* is stronger than racial prejudice in the predominantly white and economically dynamic South and Southeast. This stereotype draws on all the essential features of the Northeast: its backwardness, poverty, racial heritage, and traditional culture. The inequalities that reinforce these prejudices show few signs of diminishing in the near future.

THE SOUTHEAST

In the 1690s, the discovery of gold in the mountains and streams of southeastern Brazil triggered the shift in population, economic growth, and political power away from the Northeast to the Southeast. After 1700, the gold rush initiated the settlement and colonization of the Brazilian interior, and it made the tiny coastal settlement of Rio de Janeiro Brazil's principal port and commercial center—and its capital—by the mid-eighteenth century. In the nineteenth century, the rise of coffee cultivation around Rio de Janeiro and São Paulo consolidated the political and economic supremacy of the Southeast, which in turn intensified the contrast between that region and the declining Northeast. By the mid-nineteenth century, two Brazils had begun to emerge: the traditional poverty-stricken and economically backward Northeast, and the dynamic, cosmopolitan, industrializing Southeast. Over the past century, the contrast and the gap between these two regions has increased.

The states of Espirito Santo, Minas Gerais, Rio de Janeiro, and São Paulo account for just 11 percent of the nation's area, but they contain 43 percent (63 million) of its population. The population density of these four states (68 inhabitants per square kilometer) is more than two-and-one-half times that of the Northeast (27 inhabitants per square kilometer). The Southeast has become the most urbanized section of the country (88 percent versus 60 percent in the Northeast) and is home to Brazil's three largest cities. São Paulo (12,588,725), Rio de Janeiro (9,014,274), and Belo Horizonte (2,540,130) form an industrial triangle that produces 80 percent of the nation's manufactured goods. Greater São Paulo alone accounts for just over half of Brazil's industrial output. Consequently, Southeasterners enjoy a standard of living nearly one-and-one-half times that of the national average, and three-and-one-half times that of Northeasterners.

Culturally and ethnically, the Southeast forms a transitional zone between the more African Northeast and the European South. Blacks and mulattos comprise nearly half the population of Minas Gerais and Espirito Santo, just under 40 percent of the population of Rio de Janeiro, and just 20 percent of the population of São Paulo. Whites make up more than 80 percent of the population of the South, over 90 percent in the southernmost state of Rio Grande do Sul. In the Southeast, European and African Brazil meet and intermingle.

Although small settlements had been founded on or near the coast as early as the 1530s, intensive colonization of the coast and the interior did not get under way until the 1690s, with the discovery of gold in the mountains some 250 miles north of Rio de Janeiro. The first great gold rush in the Western world drew tens of thousands into the mountainous region that became known as General Mines (Minas Gerais). A region extraordinarily endowed in mineral wealth, Minas Gerais was once described by a French mining engineer as a "heart of gold in a breast of iron." In the eighteenth century, Brazil produced the vast majority of the gold that circulated in Europe and helped fuel England's industrial expansion. To the north of the goldfields, the Portuguese also found some of the richest diamond deposits in the world, which made Brazil the world's leading supplier of this precious stone until the late-nineteenth-century diamond rush in South Africa.

Dreams of quick wealth lured tens of thousands of Portuguese, Brazilians, and their slaves into the mining zone via two routes: down the São Francisco River from the Northeast, or overland through the mountains from Rio de Janeiro or São Paulo. The river valley became a major cattle ranching region that supplied foodstuffs to the mining zone, while Rio quickly became the major port of entry for slaves and commerce. Dozens of small mining towns sprouted up along the mountainous river valleys in central Minas Gerais, but no single major urban center arose. Vila Rica de Ouro Prêto (Rich City of Black Gold) became the capital of the captaincy, and, along with the other mining towns in the area, it went into decline as the gold played out in the last third of the century. By the early nineteenth century, gold production had plummeted to roughly one-fifth of its mid-eighteenth-century level. By the nineteenth century, Minas had evolved into a relatively isolated ranching and farming economy with a substantial intra-regional trade. All that remained of the glorious gold rush days were the beautiful baroque churches in the decaying mining towns.

Southern Minas Gerais participated in the expansion of coffee cultivation, and its economy became part of the hinterland of Rio de Janeiro and São Paulo in the last half of the nineteenth century. The core of the new industrial economy arose in the old mining zone around textiles and metallurgy. Blessed with some of the richest iron ore reserves in the world, the state, beginning in the 1930s, built the foundations of an industrial economy around iron and steel.

Belo Horizonte (Beautiful Horizon) became the focal point of this industrial expansion. Inaugurated as a planned city in 1897 to replace the isolated colonial capital in Ouro Prêto, Belo Horizonte grew at an astonishing pace in the years after World War II. A small city of some 55,000 in 1920, Greater Belo Horizonte has today become a modern metropolis of glass and steel skyscrapers, heavy industry, and a population that approaches four million. As an industrial center in a minerals-rich state, Belo Horizonte has the look and feel of Denver without the snow and extreme cold. Although Minas Gerais is located in the tropics, its mountains temper the climate, which has but two seasons. From October to April, warm, moist air from the South Atlantic pushes up into the region, as maximum temperatures approach the high 90s and rain becomes a daily occurrence. From May to September, cool temperatures (ranging from the low 40s to the low 80s) and dry weather prevail.

The second most populous state in Brazil (after São Paulo), Minas Gerais has developed its own peculiar cultural characteristics. The people (mineiros) of this landlocked, mountainous, and insular region are known for their deep family loyalties, a mistrust of outsiders, financial acumen, and conservative politics. Along with São Paulo and Rio de Janeiro, Minas dominated Brazilian politics in the late nineteenth and early twentieth century, and the state has produced some of Brazil's most famous political figures. At virtually every major political turning point in Brazil since 1700, Minas has played a crucial, if not pivotal, role.

Despite the prominent role of Minas Gerais in national politics since the eighteenth century, economically it has generally lagged behind Rio de Janeiro and São Paulo, forming a part of their hinterland. The Rio–São Paulo axis comprises the true heartland of Brazil, and has since the beginning of the century. For nearly two centuries, the city of Rio was the financial, industrial, and cultural capital of Brazil. Since the turn of the century, the city of São Paulo has gradually challenged Rio's primacy.

The shift of power and influence from Rio to São Paulo has been most dramatic during the past three decades. São Paulo replaced Rio as the economic center of the country by World War II, and it has gradually become Brazil's financial giant. Since the inauguration of Brasília in 1960, Rio has steadily lost its position as the political and administrative center of the nation. In the 1970s and 1980s, São Paulo challenged Rio's traditional role as the cultural emporium of the country, and it has become an increasingly important center for art, filmmaking, publishing, and academia. Much like New York City, Rio is a city whose era has passed, despite its continuing (and exceptional) cultural and political influence.

In spite of its relative decline, Rio de Janeiro remains South America's most beautiful and most alluring city. Located on the stunning Guanabara Bay, Rio rises up suddenly out of the beaches and lowlands, where solid basalt mountains jut up against the Atlantic. The crescent-shaped beaches at Ipanema and Copacabana, along with Sugarloaf mountain and Corcovado mountain with its monumental statue of Christ the Redeemer, have become the most famous physical symbols of Rio. As is true in southern California, geography has shaped the development, the culture, and the image of Rio de Janeiro. Sunshine and beaches have defined the stereotype of the *carioca* and southern California.

While the Southeast forms the transition zone between northeastern and southern Brazil, Rio is the point at which African and European, traditional and modern Brazil meet. If Brazil's soul is in Salvador, surely its heart is in Rio de Janeiro. With its exquisite colonial architecture, shining skyscrapers, international business set, and racially diverse inhabitants, Rio de Janeiro reflects a complex and fascinating blend of historical epochs and cultural traditions.

The Portuguese founded Rio de Janeiro (River of January) in January 1565 to counter efforts by the French to establish a foothold along the Brazilian coast. Not until the discovery of gold in Minas Gerais did the settlement attract much attention. With is strategic location and excellent harbor, Rio quickly became the major transit point for commerce and slaves going into and out of the mining zone. With the decline of sugar in the Northeast and with the gold rush in the Southeast, the Portuguese transferred the capital of their American colony to Rio de Janeiro in 1763.

The transfer of the Portuguese monarchy to Rio in 1808 transformed this city into the largest in South America (population 175,000) in the early nineteenth century, definitively establishing its cultural, economic, and political preeminence over all contenders. Gold, and later coffee, fueled the slave trade, Africanizing local society and giving Rio the largest urban slave population in the hemisphere (one-third of the city). Massive urban renewal at the turn of the century also transformed Rio into a cosmopolitan metropolis that was consciously modeled after Paris. City planners carved out broad avenues and luxuriant parks. They also leveled hills and reclaimed land from the ocean, beginning the construction of the beaches that make the city so famous.

Since the 1940s, when Brazilians from the countryside, and particularly from the Northeast, began pouring into the city in search of jobs and a better life, Rio has been transformed in ways never envisioned by planners. On the steep hillsides and the outlying lowlands of the city, dozens of slums (*favelas*) sprang up to accommodate the hundreds of thousands of migrants. Nearly a quarter of the population (some 2 million people) live in the *favelas*. A quarter of a million people live in the largest of the *favelas*, Rocinha, reputedly the largest slum in Latin America, it has a population roughly the size of Albany, New York.

The startling contrast between wealth and poverty in Brazil is nowhere more striking than Rio. One North American sociologist has described the exclusive beach area as "two blocks of Paris along the beach with Ethiopia surrounding it." Some of the city's biggest *favelas* cling to the steep hillsides that overlook the rich suburbs of Copacabana, Ipanema, and Barra da Tijuca. For decades, the *favelados* have provided a cheap labor force for Rio's industries and homes. The lowlands to the north of the city (the *baixada fluminense*) have become a huge industrial complex surrounded by slums whose population is in the hundreds of thousands.

As the years pass, many of the *favelas* have become permanent communities with access to some city services (electricity, water, sewage) and with highly developed community organizations. More recently, they have become the focal points of the drug trade (primarily cocaine) in Rio. Drug dealers have established their control of many of the *favelas* through bribery and violence. Periodically, struggles to control the drug trade and police efforts to capture the drug lords have turned these *favelas* into virtual war zones. A decade of economic crisis has also heightened the long-standing tension between rich and poor, as violent crime increased dramatically in Rio in the 1980s. Much like New York City, Rio de Janeiro faces a severe urban crisis as it attempts to maintain its infrastructure, combat rising crime, and provide basic social services to a burgeoning underclass.

Despite its mounting problems, Rio de Janeiro remains one of the great cities of the world, and its has held on to its freespirited and exuberant atmosphere. Although the pace of life in Rio is brisker than in Salvador, the *cariocas* have been stereotyped (especially by the *paulistas*) as genial, mellow, and easygoing. The tenor of the city feels something like Los Angeles—without the pretentiousness.

Carnival *(carnaval)* stands out as the most famous symbol of the city and its personality. For three days and nights preceding Ash Wednesday (usually in late February or early March), normal life in Brazil comes to a standstill as the entire country celebrates this ostensibly Catholic pre-Lenten holiday. Although the entire country takes part, it is in Rio that Carnival has reached its greatest expression—and excesses. For three nights, beginning in the late evening, and running until well past dawn, millions of *cariocas* and foreign visitors party and parade. Carnival, especially in Rio, has become both a staged public spectacle of parades, and a parallel series of parties that feature endless dancing, opulent costumes, and increasing displays of nudity. Carnival highlights the exuberance and sensuality of Brazilian culture and *carioca* life.

The sensuality of Rio de Janeiro contrasts sharply with the sobriety of São Paulo. A famous French sociologist neatly summed up the difference between the two cities, contrasting Rio de Janeiro's "natural beauty," shaped by the "hand of God," with São Paulo's "cement beauty," shaped by modern architects. Since the turn of the century, São Paulo has gradually overtaken Rio as Brazil's largest city, its industrial and financial center, and its cultural capital. (If the state of São Paulo were a country, it would have the second-largest GDP in Latin America, after that of Brazil.) Traveling south from Rio de Janeiro to São Paulo, one leaves the old Brazil and passes into the new Brazil. In many ways, this is also a journey from the Third World into the First.

São Paulo's rise as a world metropolis (it is now the third largest city in the world) has been rapid and recent, and it has strong parallels with the growth of Los Angeles. Founded by Jesuit missionaries in 1554 on the broad plateau above the coastal escarpment, São Paulo remained a small frontier settlement for three hundred years. At elevations above 2,000 feet, this fertile plateau has some of Brazil's best farmland. On the edge of the tropics, the state has a temperate climate that experiences occasional freezes and avoids the hot extremes of the North and Northeast. During the colonial period, the town served as a base for the *bandeirantes,* who explored the interior and hunted down Indian slaves for the plantations of the Northeast. These hardy *paulistas* of mixed European and Indian ancestry and culture opened up the goldfields of Minas Gerais in the last years of the seventeenth century. As late as 1880, São Paulo had a population of 35,000 (compared with Rio's 350,000).

In the late nineteenth century, coffee transformed São Paulo. From its initial boom in the Paraíba Valley to the north of Rio de Janeiro in the 1840s and 1850s, coffee cultivation expanded westward into the state of São Paulo in the 1860s and 1870s. In the final decades of the century, the coffee export economy revolutionized the city and state. The population of the city jumped from 30,000 in 1870 to 240,000 by 1900, and the population of the state from 150,000 to 2,250,000 during the same period. Roads and rail lines crisscrossed the interior of the state, integrating the region to an extent unparalleled anywhere else in Brazil.

Although the coffee planters initially reproduced the plantation system of the Northeast with its slave labor and large estates, the system quickly broke down in the face of labor shortages. The end of the Atlantic slave trade in 1850, and abolition in 1888, compelled the coffee planters of São Paulo to seek free-wage labor. They turned to an overcrowded Europe for cheap workers. Between 1886 and 1936, approximately one-and-one-half million Europeans—principally Italians, Spaniards, and Portuguese—flooded into the fields and towns of the state. In the early twentieth century, immigrants and their children made up more than half the population of São Paulo. Between the two world wars, a quarter of a million Japanese migrated to the coffee fields, thus creating the largest nucleus of Japanese outside of Japan.

Unlike gold and sugar, coffee became an engine of economic growth, creating a diverse and sophisticated economy centered around the city of São Paulo. By the turn of the century, *paulista* entrepreneurs (many of them immigrants) had laid the foundations for industrial expansion in textiles, food processing, metalworking, and other light industries. By World War II, the city had become a center for heavy industry, and in the 1950s and 1960s it emerged as the Detroit *and* Chicago of Brazil. In the suburbs of Greater São Paulo (the so-called ABC region of Santo André, São Bernardo, and São Caetano), Ford, General Motors, Volkswagen, and other manufacturers produce one million automobiles annually. In addition to its industrial primacy, the state is also Brazil's leading agricultural producer. Besides its coffee production, São Paulo has become the nation's leading producer of cotton, rice, potatoes, and sugar (having surpassed the states of the Northeast in the 1960s).

Approximately 50 percent of Brazil's industrial output comes from the Greater São Paulo region. This extraordinary industrial concentration has created the familiar problems of developed world industrial powers: air and water pollution, dense smog, and toxic waste. In the rush to catch up with industrialized nations, newly industrializing countries like Brazil have not been anxious to impose environmental regulation, fearing that it could discourage investment and growth. The city of São Paulo offers a look at what Los Angeles or London might be like with no regulation. In the 1970s, the concentration of chemical and industrial plants around Cubatão, between São Paulo and Santos, won it the dubious distinction of the most dangerous environment in the world. A gasoline pipeline explosion in Cubatão in the mid-1980s obliterated miles of surrounding *favelas* and killed hundreds. Although serious efforts are now being made to clean up the area, rapid and unchecked industrialization turned Cubatão into an environmental nightmare.

Increasing political and cultural influence have accompanied industrial and financial growth. During the First Republic, São Paulo dominated national politics as a partner with Minas Gerais in the so-called coffee-and-cream alliance. In the 1930s, the state passed Minas as the nation's most populous, and the huge population gives the city and state tremendous clout in national politics. All major political figures must court *paulista* voters to be competitive in national elections. Representing some 31 million *paulistas,* or 1 of every 5 Brazilians, anyone elected governor or mayor of São Paulo immediately becomes a political figure with national recognition.

Like the *cariocas* and *baianos,* the *paulistas* (residents of the city are called *paulistanos*) have been stereotyped for their supposed traits. While the *cariocas* are known for their sensual and fun-loving nature, the *paulistas* are supposedly practical and sober entrepreneurs. The *paulistas* often view their compatriots (especially the *cariocas* and Northeasterners) as lazy, undisciplined, and ignorant. An oft-repeated metaphor aptly sums up the *paulista* view of Brazilians in the other twenty-five states. Brazil, so the *paulista* saying goes, is a train with a locomotive pulling twenty-five empty cars, and the state of São Paulo is the locomotive. While clearly overstated, the metaphor conveys the power of São Paulo's role as the undisputed economic powerhouse in the nation.

THE SOUTH

South of São Paulo lies "another Brazil," as many have called it. The culture and character of the three southernmost states of Brazil have been forged by the European colonization that followed independence. Successive waves of immigrants from Portugal, Germany, Italy, Russia, and Poland have made the states of Paraná, Santa Catarina, and Rio Grande do Sul into a South American version of Europe. For centuries, an isolated frontier region of the Portuguese empire, the South has had the advantage of neglect. Unlike the Northeast and Southeast, the South does not bear the deep scars of plantation agriculture and slavery. Two staunchly independent social traditions emerged in the South. In the hilly uplands of Paraná and Santa Catarina, homesteading farmers (largely European immigrants) established a society much like that of the U.S. Midwest. A society of cowboys and ranchers similar to that of the American West evolved on the rolling grasslands of Rio Grande do Sul. Distinctly European and American in their origins, these two societies in southern Brazil have little in common with the Afro-Brazilian civilization of the Northeast and Southeast, as many Southerners are quick to tell you.

The states of Paraná, Santa Catarina, and Rio Grande do Sul account for just 7 percent of Brazil's land and 15 percent of its population (22 million). After the Southeast, the South is the most urbanized (74 percent) and most densely populated (38 inhabitants per square kilometer) region of the country. The per-capita income of the region ranks second only to the Southeast, and no other region surpasses the national average. European immigration has also made the South the whitest region of the country (83 percent versus 66 percent in the Southeast).

Geography and climate have also shaped the South's history. Located below the Tropic of Capricorn, the region has a temperate climate. Absent are the hot extremes of the North and Northeast. The South is the coldest region of Brazil. Freezes are not uncommon during the winter months (June to September) and on rare occasions the South receives a dusting of snow (usually at higher elevations). Beyond the coastal escarpment (which gradually runs out in the South), the hilly uplands and plateaux of Paraná and Santa Catarina spread southward and westward. Jungle-covered river valleys can be found in the western extremes of Paraná, as well as the monumental Iguaçu Falls on the border with Argentina. Rolling grasslands that extend out of the Argentine and Uruguayan *pampas* extend across Rio Grande do Sul.

During the colonial period, southern Brazil straddled the poorly defined frontier between the Spanish and Portuguese empires in South America. For centuries, both monarchies claimed the sparsely populated territory, and it changed hands on numerous occasions. The Portuguese made several attempts to man a fortified garrison at Colônia do Sacramento (west of present-day Montevideo). In 1750, the Treaty of Madrid between the two Iberian powers demarcated the division that roughly coincides with Brazil's current borders. Spain recognized Portuguese sovereignty over the western Amazon in exchange for control of the Jesuit missions in Paraguay and the area known as the Cisplatine around the old settlement at Colônia do Sacramento. In the 1820s, to reestablish control over the Cisplatine, Pedro I embarked on an unpopular and fruitless war with Argentina that ended (through British mediation) with the creation of the "buffer" state that today is Uruguay. Historically and culturally, southern Brazil has been more closely linked to Argentina and Uruguay than to northeastern Brazil.

Beginning in the early nineteenth century, successive waves of European immigration populated this frontier and sharpened the contrasts with the states to the north. In the colonial period, the Portuguese monarchy made several attempts to populate the region with families from the Azores. Between 1824 and 1859, the Brazilian monarchy offered land to thousands of Germans who were settling in Rio Grande do Sul. After the 1850s, thousands more German colonists built the towns of Blumenau and Joinville in Santa Catarina. In the last half of the century, thousands of Italians immigrated to Rio Grande do Sul, and Poles and Russians to Paraná. Between the two world wars, large numbers of Japanese families moved into the newly opened coffee fields in the interior of Paraná.

European immigrants, especially German immigrants, have created a culture that has little in common with much of the rest of Brazil. The cities of the South are newer, more orderly, and more European in climate and design than the older cities to the north of São Paulo. Curitiba, the capital of Paraná, has often been singled out as the best planned and administered city in the country. At an elevation of 3,000 feet and slightly inland, Curitiba has the look and feel of a Western European metropolis. It is perhaps the best managed city in Brazil, with a well designed transportation system and recycling programs much like those in Europe and the United States. In the South, the neat and orderly wooden and masonry houses with sloping roofs seem a world apart from the adobe and thatch huts of the interior and the Northeast.

German influence remains especially pronounced. Cultural ties with Germany have been reinforced for decades through German-language newspapers, schools, and clubs, and through intermarriage. As late as the 1950s, Germans staffed schools for the immigrants and their descendants with a curriculum that ignored national educational regulations. (In Teutônia, Rio Grande do Sul, for example, the majority of the townspeople are of German descent, and German is still the language of choice at home. Teutônia is a center for dairy production, and the income per capita is over $8,000—three times the national average.) While other Brazilians often look upon the Southerners with a sense of awe at their success and cultural accomplishments, Southerners also tend to look down upon Brazilians from other regions of the country. There is more than a little racism in this superior attitude.

While small farmers in Paraná and Santa Catarina staked out homesteads in a tradition much like that of the American Midwest, the prairies of Rio Grande do Sul became the home of the *gaúcho* (cowboy) and the *estância* (ranch). As the French sociologist Roger Bastide astutely pointed out, while the northeastern coast was a civilization built on sugar, the *sertão* revolved around leather, Minas around gold, and São Paulo around coffee, the horse defined civilization in Rio Grande do Sul.

As on the plains of North America, Europeans brought horses and cattle to the *pampas,* where they thrived and multiplied. On these plains, which stretch from the southern Andes to the uplands of southern Brazil, the *gaúcho* evolved as the classic South American cowboy. Fiercely independent, like his North American counterpart, the Brazilian *gaúcho*—with his broad-brimmed hat, baggy pants, large belt, accordion-like boots, and enormous knife—roamed the plains on horseback herding cattle for ranchers. Eating *charque* (dried beef, corrupted into English as "jerky") and *churrasco* (Brazilian barbeque), and drinking a strong tea *(mate)* from a small gourd, the *gaúcho* became the classic figure of the South. Like the North American cowboy, the *gaúcho* has gradually disappeared with the rise of fenced plains and urban life. Nevertheless, this colorful figure dominates the folklore and traditions of the South as much as the cowboy and his culture continue to live on (in a modern, sanitized version) in the North American West.

Southerners are quick to emphasize the differences between the *gaúcho* and the *vaqueiro* (literally, cowboy) of the Northeast. They compare the former's fierce individuality, independence, and mobility to the latter's squalid living conditions and submissive subordination to the Northeast's powerful landowners. Just as the image of the fiercely independent cowboy remains a dominant symbol in many states of the North American West, the inhabitants of Rio Grande do Sul define themselves by their *gaúcho* heritage (and the inhabitants of the state are referred to as *gaúchos*).

In many ways, Rio Grande do Sul is the Texas of Brazil, a former frontier of the Spanish empire, settled by immigrants, with a cattle-ranching culture and a stubbornly independent view of themselves as at once citizens of the nation yet separate and unique. Much like Texans, the *gaúchos* have often bucked national political control and tried to go their own way. In the 1830s and 1840s (while Texas was gaining its independence and becoming a republic), the *gaúchos* fought an unsuccessful ten-year war for independence from Brazil. These Southerners have also cultivated a military tradition and, since the nineteenth century, have supplied Brazil with some of its most important military leaders. As a stepchild in the coffee-and-cream alliance, Rio Grande do Sul oscillated between cooperation with São Paulo and Minas Gerais, and open rebellion. Since 1930, the South has produced an inordinate share of Brazil's presidents, including Getúlio Vargas and João Goulart (both *gaúchos*), and three of the five military presidents of the period between 1964 and 1985.

Frontier traditions and European colonization have shaped the stereotype of the Southerner. For other Brazilians, the Southerner (especially the *gaúcho*) is aggressive, boisterous, belligerent, and often crude. Southerners think of themselves as proud, energetic, and productive—the envy of the unambitious and inefficient non-Southerner. Unlike most Brazilians, the Southerners (and the *paulistas*) see themselves as more European than American. This is truly "another Brazil."

THE NORTH

The Northeast, Southeast, and South represent 87 percent of Brazil's population, yet cover just 36 percent of the land. The other two regions—the North and the Center-West—cover two-thirds of the nation's land and account for a mere 13 percent of the population. This vast, sparsely inhabited region is the world's last great frontier, and the Amazon basin covers the bulk of it. After traveling through the region at the beginning of the century, the Brazilian writer Euclides da Cunha described Amazonia as the "last unwritten page of Genesis." The North and Amazonia roughly coincide, covering an area seven times the size of France. At the mouth of the Amazon, Marajó Island alone is larger than Switzerland. For five centuries, the Amazon has fascinated and perplexed the outsider and attracted explorers and adventurers. Without a doubt, it is the best known and most widely written about area of Brazil.

The enormous states of Amazonas (in the west) and Pará (in the east) cover most of the North and Amazonia. The region also includes the small states of Amapá (on Brazil's northernmost coast) and Roraima (sandwiched between Amazonas and Venezuela). The states of Acre and Rondônia on the Peruvian and Bolivian borders make up the remainder of the region. Just under 7 percent of Brazilians (10 million) live in the North, whose population density is 3 inhabitants per square kilometer. The vast majority of the region is uninhabited. Two major cities, with more than a million inhabitants each, account for half of the population of the North—Belém (at the mouth of the Amazon) and Manaus (1,000 miles upriver). Three-quarters of this sparse population are the descendants of both Europeans and Indians (*caboclos*).

Since the arrival of the first Europeans in the sixteenth century, the Amazonian North has been alternately portrayed as a hell or a paradise on earth. The pessimists see nothing but dense jungle, unbearable heat and humidity, hundreds of deadly predators, and tropical diseases. The optimists see a tropical Garden of Eden, vast resources with enormous potential, and seemingly unlimited land. For five centuries, Amazonia has attracted (and often swallowed up) countless outsiders in search of El Dorado, the Garden of Eden, or a world they believed to be uncorrupted by modern civilization.

Not all of the North is tropical rainforest, although the region and the Amazon basin do roughly coincide. Along the border with Venezuela and Guyana to the north, imposing mountains rise up several thousand feet on the Guyana Shield. This isolated region is the naturalist's wonderland that served as the setting for Arthur Conan Doyle's *Lost World*. To the west, along the Colombian and Venezuelan borders the jungle gives way to rolling grasslands, and savannah occasionally breaks the monotony of the Amazonian rainforest. But the 25,000 miles of rivers in the Amazon system crisscross the bulk of the North.

Called the River-Sea by the early Spanish and Portuguese explorers, the Amazon is one of the great natural wonders of the world. Stretching more than 3,900 miles from its source in the Peruvian Andes until it empties into the Atlantic, the Amazon is just a few miles shorter than the Nile. After descending from its source in the Andes (at an altitude above 18,000 feet), the river drops less than two inches per mile over the next 3,500 miles to the Atlantic. Even more impressive than the river's length is its volume and size. Because the Amazon is more than fifteen miles across in some spots, its opposite bank is sometimes invisible below the horizon. Reaching a depth of 250 feet in some places, it is the deepest river in the world. The Amazon discharges 160,000 cubic meters of fresh water *per second* into the Atlantic—about 15 percent of all the fresh water emptied each day into the world's oceans. Its daily discharge could supply *all* U.S. homes with their water needs for five months. (As a point of comparison, the Mississippi passes less than 20,000 cubic meters per second into the Gulf of Mexico.) This immense river system drains one-fifth of the world's forest.

Despite the familiar image, Amazonia is not the hottest region of Brazil. While maximum temperatures in the backlands of the Northeast range above 100 degrees Fahrenheit year round, the average temperature in Manaus hovers between 69 and 98 degrees, with an average of 80 degrees Fahrenheit. The problem in the North is the combination of high temperatures and high humidity. Although the rainy season lasts from November to June, the average relative humidity in Manaus hovers around 80 percent—year-round. The two seasons (wet/winter and dry/ summer) also produce seasonal flooding, which raises the level of the river as much as 10-20 feet during the rainy season.

Amazonia is a naturalist's paradise. It is also one of the world's great biological frontiers. According to some estimates, less than half of all Amazonian flora has been identified and only two-thirds of the fish species have been named. The renowned sociobiologist, Edward O. Wilson, has described the "unsolved mysteries of the rainforest" as ". . . formless and seductive. They are like unnamed islands hidden in the blank spaces of old maps, like dark shapes glimpsed descending the far wall of a reef into the abyss." Tom Lovejoy of the Smithsonian Institution has calculated that the Amazon basin contains some 10 percent of all plants and animals on the face of the Earth. Amazonia is an enormous uncharted genetic reservoir of biodiversity.

The first Europeans to navigate this immense river also unwittingly gave it a name. In 1541, fresh from the conquest of the Inca empire in Peru, a Spanish expedition crossed the towering Andes and descended into the tropical jungle in search of the ever-elusive El Dorado. Gonzalo Pizarro, a brother of the conqueror of Peru, led the ill-fated expedition that began to disintegrate in the jungle and pitted jealous and ambitious adventurers against each other. On a makeshift craft, and led by Francisco Orellana, half of these desperate men set off down river, ostensibly in search of supplies and help for Pizarro and the other remaining men. Whether by design or natural constraints, the group never returned. Over the next year, they survived Indian attacks, disease, starvation, and bitter internal squabbles as they traversed the length of the mighty river, eventually returning to Spanish civilization along the northern coast of South America.

Father Gaspar de Carvajal left a chronicle of this extraordinary journey. In one section of his account, Carvajal describes a furious battle with women warriors likened to the ferocious Amazons of Greek mythology. "We ourselves," he claims, "saw these women who were there fighting in front of all the Indian men as women captains, and these latter fought so courageously that the Indian men did not dare to turn their backs, and anyone who did turn his back they killed with clubs right there before us. . . . These women are very white and tall, and have hair very long and braided wound about the head, and they are very robust and go about naked, but with their privy parts covered, with their bows and arrows in their hands, doing as much fighting as ten Indian men." In the last four centuries, no one has been able to find Carvajal's vaunted warriors on what became known as the River of the Amazons.

Although the Spanish, coming in the hard way over the Andes, were the first Europeans to navigate the great river, the Portuguese claimed the region. More interested in the sugar plantations of the Northeast and the goldmines of Minas Gerais, the Portuguese paid little attention to Amazonia and the North. In 1616, they did establish a fortified garrison at Belém (Bethlehem) at the mouth of the Amazon, largely to fight off French efforts to establish a colony on the northern coast of South America. For centuries, Catholic missionaries and slave traders vied for control of the Indian population. Few whites entered the region until the nineteenth century.

The Portuguese monarchy and its Brazilian successors have long feared that other European powers or the United States would try to wrest Amazonia and its riches from them. In the mid-nineteenth century, only after great pressure from the United States and Great Britain did the Brazilian monarchy open up the river to international shipping. In the 1850s, Brazilian steamships began to carry trade and passengers between Belém and Manaus. The region remained unexplored and unexploited by Brazilians and foreigners alike until the last decades of the century.

In the late nineteenth century, an exploding demand for rubber in the industrial nations transformed the Amazon into a booming economy. As the automotive industry expanded in Europe and the United States, manufacturers eagerly sought natural rubber for tires. Natural rubber could be found in just two places—Central America and Brazil—and *hevea brasiliensis* quickly proved to be the superior product. Rubber trees are plentiful yet widely dispersed in the rainforest, and an elaborate, primitive, and exploitative system was developed to get the rubber from forest to factory. In the forest, *seringueiros* (from a Portuguese word for rubber, *seringa*) located and regularly tapped the milky latex from trees. Middlemen traveled up and down the river, trading goods for rubber and often keeping the *seringueiros* in debt. Known as *aviadores* (lenders), these middlemen then sold their rubber stock in the cities for export, largely to England.

Life in the forest was often "nasty, brutish and short," and the great beneficiaries of the boom were the exporters, who made fortunes very quickly. The Brazilian "rubber barons" turned the sleepy and squalid city of Manaus into a modern European city in the tropics. At the confluence of the Solimões and Negro Rivers, Manaus more than doubled in size, going from a population of 29,000 in 1872 to over 75,000 by 1920. (Belém grew at an even faster pace, going from 61,000 to 236,000 in the same time period. The population of the states of Pará and Amazonas also quadrupled in size.) The rubber barons spent lavishly—constructing enormous mansions, importing European cobblestones to pave streets, and building an opera house that would have been a jewel for any European city. Some of the greatest performers of the era received fabulous sums to perform for the *nouveaux riches* in what must have been the only tropical opera house in the world. British engineers designed and built for the ocean-going vessels a special floating dock that connected Manaus to the North Atlantic economies.

Like the sugar and gold cycles, the rubber boom rose and fell, but at a much faster rate. Between 1880 and 1914, Brazilian rubber production rose constantly while the country held a virtual monopoly on production. The rise of new producers in Asia shattered the Brazilian monopoly and sent the Amazonian economy into a tailspin from which it took decades to recover. For nearly a century, an old legend has been perpetuated about the end of the rubber bonanza. Supposedly, Henry Wickham (later Sir Henry for his work) smuggled rubber seedlings out of the Amazon in 1876. After careful cultivation in England's Kew Gardens, the British transplanted the seedlings to Ceylon and Malaya. They quickly developed enormous plantations at the turn of the century. Cheap Asian labor, production on a concentrated and massive scale, and British entrepreneurship then drove Brazilian rubber from the marketplace.

For decades, critics have taken the Brazilians to task for failing to respond to the British challenge with modern plantation-production techniques. Brazilians, on the other hand, have repeatedly denounced unfair economic practices and "imperialism" for the demise of Brazilian rubber. In fact, Wickham did take the seeds, although he embellished his account with the drama of smuggling when he simply shipped the seedlings to Britain. Brazil's failure, more importantly, had little to do with business practices and entrepreneurship. Brazilians and foreigners alike have failed for decades to develop rubber plantations for the simple reason that it has been ecologically impossible. When the dispersed trees are concentrated on plantations, an indigenous leaf blight *(microcyclus ulei)* appears and kills all the *hevea* plants. South American leaf blight has defeated the efforts of many Brazilian and foreign entrepreneurs (including Henry Ford) to produce plantation rubber. The disease has yet to appear in Asia or Africa. A fungus—and not economics or human failings—destroyed the Brazilian rubber boom.

By the 1920s, the Amazonian economy had entered into a long-term decline that was reinforced by Asian plantation rubber and the invention of synthetic substitutes. Although Brazilians (especially Northeasterners) continued to move into the region, the North's economy stagnated until the late 1960s brought the military to power. Like the Portuguese and Brazilian rulers before them, the generals feared intervention in the Amazon basin by powers intent on weakening Brazil and exploiting the region's vast resources. The generals also saw Amazonia as central to their plans to make Brazil into a world power. They thought the region's untapped resources would fuel a drive for industrialization and economic growth. The vast open lands would defuse the pressure on land in the crowded Northeast. Impoverished and unproductive peasants in the Northeast would become productive pioneers in the Amazon. As one military president put it, the systematic colonization of the region would unite men without land in the Northeast, and land without men in the Amazon.

In the 1970s, the military regime promoted a sweeping plan to open up the North on a scale hitherto unimagined. The first task of "Operation Amazon" and the "Plan for National Integration" was to open up highways across a region where airplanes and boats offered the only viable transport. In the seventies, the government forged ahead with the construction of the 11,000-mile-long Trans-Amazon Highway, which began at Recife and João Pessoa on the northeastern coast; crossed the *sertão;* and ran east-to-west across southern Pará and Amazonas, and through Acre to the Peruvian border. Three major north-south highways would connect up the industrial South and Southeast via Brasília. These roads were to serve as channels of growth, bringing settlers and commerce to the North.

The military government also undertook a systematic aerial survey of the North in search of natural resources. They were soon rewarded. In 1967, geologists discovered an immense iron ore deposit at the Serra dos Carajás in southern Pará. Along with major multinational corporations from the United States, Japan, and Europe, the military began to develop Carajás and major bauxite, uranium, and tin reserves in the Amazon basin. The export-led growth model of the military regime made the raw materials of the Amazon vital elements in economic growth. The Carajás complex is nearing completion and will soon be the largest iron-mining operation in the world. A new railway connects this isolated operation with ports on the coast, and nearby the government is nearing completion of one of the largest hydroelectric dams in the world to produce power for the complex.

The rush of immigrants into the region touched off a new gold rush in southern Pará. In 1980, prospectors stumbled across an enormous gold deposit at Serra Pelada. Within two years, a ram-shackled mining town of more than 100,000 had sprouted up around an ever-widening crater worked by 50,000 miners. In scenes reminiscent of nineteenth-century gold rushes, violence, prostitution, and smuggling forced the authorities to step in to establish order. The government placed the area around Serra Pelada under military control and set up a system of claim-staking and gold purchasing. Men made and lost fortunes overnight. Gold production, for the first time since the early nineteenth century, ensured that Brazil was pushed back into the ranks of the major producers. By the late 1980s, Serra Pelada's gold was exhausted, and the huge open pit mine had become a lake. Tens of thousands of small-time prospectors, however, have fanned out across the entire Amazonian basin and are dredging rivers and streams for placer gold.

Along with exploitation of resources, colonization became the other major objective of the military's drive to develop the Amazon. The Brazilian Institute of Agrarian Reform and Colonization (or INCRA, its acronym in Portuguese) drew up ambitious plans to hand out land with a system similar to that mapped out by the U.S. Homestead Act in the nineteenth century. Tens of thousands of landless peasants, dreaming of buying their own small farm, streamed into the North. The new highways, especially along Highway BR-364 through Mato Grosso and Rondônia in the West, channeled the flow of migrants. Along these new routes, peasants burned off the jungle and cleared their small plots.

Despite grandiose plans to distribute land to the poor peasants of the Northeast, the government handed over the vast majority of land to large landowners, many of whom already had huge holdings in the

Southeast and Northeast. The military regime's fiscal and tax policies attracted investors who could make money without creating viable and productive cattle ranches. Many ranchers simply established the bare infrastructure of a cattle ranch to qualify for the lucrative government subsidy programs. (One *paulista*, Pedro Dotto, controls more than 5 million acres—an area the size of El Salvador—near the Peruvian border.) In a rerun of the Old West, large ranches sprang up, employing their own hired hands, and their own hired guns to enforce the law of the powerful on a chaotic frontier.

The poor tropical soils of the North quickly frustrated the dreams of countless migrants and stymied the plans of government technocrats. It took a short time to discover that these soils were inadequate for cattle ranching and farming. The poverty of the soils drove small and large holders alike to burn and clear larger and larger areas, thus compounding a growing ecological crisis. By 1990, deforestation had destroyed about 10 to 15 percent of the Brazilian Amazon (although the rest has been largely untouched). The hundreds of thousands of fires also began to have a serious impact on carbon dioxide levels in the atmosphere.

According to the estimates of scientists at the Goddard Space Flight Center who tracked the western Amazon via satellite in 1987, nearly 8,000 fires burned daily. They calculated that, during the burning season, nearly a quarter million fires sent more than 10 million metric tons of particulant matter into the atmosphere. In the 1980s, ranchers and miners leveled the forest at the rate of more than 50,000 square miles a year. By 1985, the dramatic expansion of deforestation and burning heightened fears that the destruction of the Amazon rainforest would have long-term and irreversible effects on the planet. While no one knows what the impact of global warming and the greenhouse effect will be, the massive deforestation of the Amazon in the 1980s focused international attention on Brazilians' move to the North.

As the international outcry over the Amazon's devastation intensified, the sensitivity of Brazilians to outside interference increased. For them, especially for the government, criticism seemed hypocritical when it came from nations that historically have, and currently do, clear their own forests and produce the overwhelming bulk of atmospheric pollution. (The Goddard study also showed that logging had damaged the forests of the Pacific Northwest more severely than the Brazilian Amazon. A mere 10 percent of the original forest in the Pacific

Northwest remains.) The Brazilians, logically enough, often suspect that the industrial nations are more motivated by protecting their own economic supremacy than by ecological and humanitarian concerns. They see the path the industrial nations have taken, and the environmental devastation these countries have produced, and ask: "Who are you to criticize us?" It is a question we should confront honestly and directly before we judge the Brazilians too quickly.

While the opening of the North may ultimately produce enormous ecological damage, the Brazilian Indians are currently paying a greater and more immediate human price than any other group. For centuries, the Indians of the Amazon have been relatively successful at surviving and avoiding the fate of their precolumbian neighbors, who lived on the plains and in the forests of North America. Perhaps as many as a million Indians survived into the twentieth century in the Amazon basin, largely due to their isolation. The opening of the North irrevocably changed their fate.

For decades, Brazilian Indians have been considered wards of the state and set aside on enormous reservations in the interior, with their affairs controlled by a government agency. In 1910, a group of idealistic young army officers helped create the Indian Protection Service (SPI) to protect the tribes of the interior. Led by Colonel Cândido da Silva Rondon, who carried out a number of scientific and military expeditions into the interior at the turn of the century, the SPI attempted to protect the Indians through legislation that guaranteed their rights to their tribal lands. With a pacifistic approach ("Die if necessary, but never kill") the SPI created sixty-seven Indian reserves, mostly in the North and Center-West.

The slow but steady entry of whites into the interior, however, continued to put pressure on the Indians. By one estimate, contact with whites deculturated or destroyed more than eighty tribes between 1900 and 1957. In the same period, the indigenous population of Brazil dropped from an estimated 1 to 3 million, to less than 250,000. In the 1960s, more than 120 tribes, mostly in groups of 100 to 500, continued to live in the Amazon and survive off hunting, fishing, and gathering.

By the 1960s, the SPI had become a corrupt bureaucracy that did great damage to the Indians. In response to enormous international protest, the military dismantled the SPI in the late sixties and replaced it with the National Indian Foundation (FUNAI). The main objectives of the FUNAI have been to integrate Indians into national life and guarantee that they do not stand in the way of efforts to develop the Amazon. The drive to exploit the forests, mineral deposits, and waterways of the North has placed enormous pressures on the Indians as their reservations have been invaded or moved, and their way of life threatened. They are, as the anthropologist Shelton Davis has put it, "victims of the miracle."

Although the economic crisis of the eighties and the return to civilian politics slowed the original development plans of the generals, they unleashed forces that will not be easily contained. The desire of Brazilians to make their nation into a power seems to require the opening and exploitation of the North. It now appears that the process may be irreversible. Without doubt, the development of the Amazon cannot be only a Brazilian issue, but must be a global one. Over the next few decades, the Brazilian Indians will pay a heavy price for the "development" of the Amazon. The price to be paid by the rest of us for the opening of the world's last great frontier remains to be seen.

THE CENTER-WEST

Surpassed only by the North in size and remoteness, the Center-West makes up the rest of Brazil's vast frontier. A region that has traditionally been even more isolated than the Amazon, it has, in the last four decades, become the fastest growing region in Brazil. The construction of Brasília in the late fifties, and new highways in the seventies, has placed the Center-West directly in the path of Brazil's future.

Until the 1950s, just two states, Goiás and Mato Grosso, formed the Center-West region. The construction of the Federal District on the Goiás–Minas Gerais border and the creation of the new state of Mato Grosso do Sul in the 1980s reflect the economic and demographic growth in the Center-West in the past four decades. These three states and the Federal District cover 18 percent of Brazil's territory but contain just 6 percent of the nation's population. Much like the North, the Center-West has a very low population density (6 inhabitants per square kilometer) and vast unpopulated tracts. Ethnically, the population has traditionally been much like that of the North; a mixture of Europeans and Indians (*caboclos*), and pockets of Indians on reservations.

Amazonia, with its characteristic dense tropical forest, extends across much of the north and west of the region. Sub-tropical forest, scrub brush, and grasslands cover most of the south and east on a vast sloping plateau. This plateau rises upward from elevations around 1,000 feet in the south to nearly 4,000 feet around Brasília in the east. Elevation offsets latitude and produces a moderate tropical climate with a cool, dry season and a wet, warm one. On the western fringes of Mato Grosso the world's largest swamp, the Pantanal, stretches across nearly 8,000 square miles of hot, humid tropical terrain during the rainy season (January to April).

As in the North, there have been pockets of settlement in the Center-West since the early seventeenth century. A large river system also provided the only viable means of travel into and out of the region. For centuries, the easiest means of traveling from Rio de Janeiro to the interior of the Center-West was via the La Plata and then up the Paraná River. The highlands and deep river valleys of the region made overland travel extraordinarily demanding.

European settlement and colonization of the Center-West began in the early eighteenth century with the gold rush in Minas Gerais. A smaller rush pulled prospectors and miners westward from Minas into Goiás and Mato Grosso. An extensive system of Jesuit missions lay on the poorly defined and hotly disputed frontier between the western rim of the Portuguese empire and the eastern edge of the Spanish empire in South America. In 1750, the Portuguese renounced all claims to the mission territory and tentatively demarcated the western border of Brazil with modern-day Paraguay. For the next two centuries, farmers, ranchers, and prospectors formed small settlements on this isolated frontier. As late as the 1940s, no major roads connected the South and Southeast with Mato Grosso and Goiás.

Life in the Center-West changed dramatically with the construction of Brasília in the 1950s. Since at the least the eighteenth century, Brazilians had discussed plans to build a new capital in the interior of the country. The Constitution of 1891 even outlined the legislative mechanisms for the move. In the 1940s and 1950s, several commissions studied possible sites in Goiás and Minas Gerais. The election of Juscelino Kubitschek to the presidency in late 1955 changed centuries of planning and discussion into action.

In his campaign, Kubitschek pledged to build Brasília—and he kept his word. He envisioned the new capital as the symbol and catalyst of a dynamic, modern Brazil. It would be, as he put it, a "capital of hope." A technical commission selected a site in eastern Goiás on a broad plateau near the Minas Gerais border. With people working around the clock, Brasília took shape on this uninhabited savannah in the late fifties. Kubitschek often left Rio de Janeiro in the evening, flew to Brasília, and spent the night supervising the project before flying back to Rio at dawn. (The construction of Brasília also made a number of builders and planners very rich, a phenomenon that generated persistent accusations of corruption and political favoritism.)

Lúcio Costa, Oscar Niemeyer, and Roberto Burle Marx, all internationally recognized architects, designed the new capital. Devoted students of the legendary Swiss architect Le Corbusier, Costa and Niemeyer produced designs that were rational to a fault. They laid out the city on two broad axes in the shape of an airplane with slightly bent back wings. Along these wings they placed the residential areas, commercial centers, and hotels. Along the fuselage were government offices. The Plaza of Three Powers sat in the cockpit with the presidential palace, supreme court, and legislature facing each other across the plaza. Down each side of the forward area of the fuselage, modernistic office buildings housed the cabinet ministries. A dam created a large, artificial lake in front of the plane and provided a beautiful locale for the homes of high government officials and foreign embassies.

Costa, Niemeyer, and Marx designed ultra-modernistic sculptures, gardens, and buildings. When it was inaugurated in April 1960, Brasília seemed to many to represent a rational, modernistic style, the perfect symbol for Kubitschek's vision of the new Brazil. For its critics, the city has always seemed cold, artificial, and overplanned—the dream of technocratic planners, and the nightmare of the humanist. The art critic Robert Hughes has called it the ultimate symbol of the bankruptcy of the modernist style. Critics and supporters alike agree on one thing: the construction of Brasília was an extraordinary accomplishment, and without Juscelino Kubitschek it probably would not have been possible.

The once-uninhabited savannah now boasts a city of more than a million. Growing even faster than its most optimistic boosters anticipated, the city has already overrun the rational design of its architects. The government bureaucracy has outgrown its glass-box ministry buildings, and housing seems perpetually in short supply. On the periphery of the city, shantytowns of immigrants from the Northeast have mushroomed, and these "satellite" cities house thousands of workers who commute long distances to work in the capital.

Despite its size and rapid growth, Brasília remains an eerie place. Built with broad avenues across great distances, the city has forced the lower classes to commute long distances. Planners envisioned a twenty-first-century metropolis moved by automobiles. In a country where only about a quarter of the population can afford a car, the reality has been a city of long-distance bus rides. The broad avenues and relative scarcity of automobiles have produced a city with an empty feeling at its core. Brasília is a cultural backwater, offering little in the way of theater, dance, and music. São Paulo and Rio de Janeiro continue to dominate the cultural life of the nation. This eminently bureaucratic city ebbs and flows with the pulse of government. On weekends, politicians desert the capital in droves, preferring to spend their time elsewhere. The capital holds little attraction for the foreign diplomatic corps either. Many embassies resisted the move from the cosmopolitan Rio. The French were the last to move their embassy, not making the transfer until 1974. It will take decades, no doubt, for this ultra-rational creation to take on a more human feel, and to develop the amenities of older, less carefully designed cities.

The transfer of the nation's capital has succeeded in opening up the Brazilian interior. Acting as a magnet and a pole of growth, Brasília has drawn hundreds of thousands to the Center-West, largely from the drought-stricken Northeast. Northeasterners built Brasília, they dominate its slums, and they continue to flow across the region and into the Amazon basin. In addition to Brasília, the major growth areas have been northern Goiás (which became the state of Tocantins in 1988); the new state of Mato Grosso do Sul; and the western fringe of Mato Grosso, which borders on Rondônia.

All three areas have experienced a farming and ranching boom during the past two decades. Mato Grosso do Sul (along with bordering western Paraná) has become the fastest-growing producer of coffee and soybeans. Large ranching interests dominate Tocantins, and its economy has been spurred by the mining boom in neighboring southeastern Pará. Northwestern Mato Grosso forms part of Amazonia and has participated in the boom created by the government's drive to open the frontier.

The construction of the Trans-Amazon Highway and its north-south feeder highways has accelerated and facilitated the economic boom and the massive movement into the Center-West. All the major highways that crisscross the region feed into Brasília, just as all the major highways from the developed Southeast feed into the capital. Brasília has been the stimulus for the movement into the Center-West, and a catalyst in the full integration of the transport network in the industrial triangle of the Southeast (São Paulo, Belo Horizonte, Rio de Janeiro). Juscelino Kubitschek's dream of turning Brazil's attention inward has succeeded. The construction of Brasília in the 1950s signalled the opening of Brazil's frontier. Just as North Americans witnessed the closing of their frontier in the 1890s, the world may witness the closing of the last great American frontier in the 1990s.

BRAZIL, BRAZILS

Comparing the closing of these great American frontiers highlights the extraordinary speed of recent Brazilian expansion. The United States began its westward expansion as a young nation, fresh from a war for independence. The Brazilians began the westward expansion into their frontier in the 1950s, after a century-and-a-quarter of nationhood. The closing of the U.S. frontier stretched across a century, from the crossing of the Appalachians to the end of the Indian wars in the West. Although the frontier expansion in Brazil has taken centuries to initiate, it might well take no more than a few decades to complete. As Alex Shoumatoff has noted, it is a "high-tech" frontier, and expansion moves at a pace unimaginable in the nineteenth century. Aircraft, telecommunications, and the automobile have accelerated the westward movement. Brazil is making up for its late start with a more intense and rapid expansion.

European-Americans moved gradually and relentlessly across the North American frontier, from coast to coast, from the eighteenth century to the twentieth. In the nineteenth century, the cultural and technological disjuncture between Western and native civilizations produced a bloody and costly clash on the U.S. frontier. The Brazilians, on the other hand, have moved quickly and fitfully across their frontier in a few decades. This cultural and technological disjuncture between Native Americans and Westerners has been more severe, and the clash more devastating.

Brazil's rapid frontier expansion has vividly revealed the nation's diversity by bringing the many Brazils face to face with each other. The drive to develop the nation has produced better transportation and communications, thus connecting the nation's diverse regions. Migrants from the old plantation Northeast or the arid *sertão* flock to the industrial cities of the Southeast, the European South, and the jungles and savannahs of the North and Center-West. Meanwhile, the economies and people of the South and Southeast have pushed into the Center-West in search of opportunity.

On the dry plains of central Brazil, the diverse regions of the nation converge, and contrasts stand out. Traveling from the Indian reservations in the Amazon basin to ultra-modern Brasília, one traverses 500 years of Brazilian history and regional diversification. Brazil has turned inward on itself confronting the contrasts of its past and present. Brazil began as an Africanized plantation society in the Northeast. Gold, coffee, and then industry created a new Brazil in the Southeast, one that gradually spun off a European Brazil in the South. These newer Brazils have increasingly diverged from the old Northeast, the traditional life of the arid backlands, and the rudimentary civilization of Indians and pioneers on the frontier of the North and Center-West.

Brazil's extraordinary regional diversity has developed over centuries and will not fade in the foreseeable future. This regionalism makes Brazil a complex and fascinating country. Like other nations of continental dimensions, Brazil is really several countries within a country. One of the keys to the success of these nations has been their ability to reconcile the interests and unique characteristics of these smaller "countries" with a larger national interest. It has been neither easy nor painless, and the process of reconciliation continues. If Brazil is to achieve its potential and to become the country of the future, it must find ways to reconcile the strengths and weaknesses of its diverse regions. For several centuries now, the Northeast has borne the heavy burdens of Brazil's past, while the Southeast and South have held the key to its present. The integration of these regions with the North and Center-West will determine Brazil's future.

THREE

Lusotropical Civilization

> The formation of Brazilian society . . . has been
> in reality a process of balancing antagonisms.
> Economic and cultural antagonisms. Antago-
> nisms between European culture and native cul-
> ture. Between the African and the native.
> —Gilberto Freyre

BRAZILIAN SOCIETY HAS BEEN FORGED out of a centuries-long clash
between peoples and cultures from four continents: Europe, Africa,
Asia, and the Americas. Five centuries of complex interaction among
these peoples and cultures has produced a unique society that is at once
both Western and non-Western. The interpenetration of features from
cultures on four continents has also created a society with stark contrasts
as well as extraordinary racial and cultural mixture. Perhaps nowhere else
in the world is the contrast between rich and poor so striking, the
continuum of skin colors so diverse, and the blend of Western and non-
Western features so subtle.

Since the 1920s, Brazilian intellectuals have highlighted the extraordinary mixture that is Brazil. Gilberto Freyre, Brazil's most famous
"social scientist," even argued forcefully in his voluminous writings that
Brazil represented a unique civilization in world history, a "lusotropical"
civilization. He believed that this mixture of Portuguese (Lusitania to
the Romans, hence the "Luso") and tropical cultures made Brazil
uniquely capable of bridging the chasm between the European industrial
societies of the Northern Hemisphere and the non-white agrarian
societies of the Southern Hemisphere. For Freyre, Brazilians could take

great pride in their racially and culturally mixed heritage as an example for other nations and societies to follow.

Although Freyre has been justly criticized for painting an unrealistic portrait of a Brazilian "racial democracy," he quite rightly recognized that racial and cultural mixture defined and shaped Brazilian society and culture. Most Brazilian intellectuals in the nineteenth century tried to ignore this mixture or they bemoaned it while grappling with ways to Europeanize their society. Since the 1930s, and to a large extent due to the pioneering works of Freyre, racial and cultural mixture has become universally recognized as a defining—if not *the* defining—feature of Brazilian society.

Although it is difficult to generalize about such an elusive thing as national characteristics, certain features do seem to define a national culture. As with all large societies, these features do not apply to everyone, and they can be endlessly qualified with exceptions. The complexity of Brazilian society and culture challenges the bold generalizations that I will make in the following pages. Yet I cannot back away from making generalizations about national patterns and attempting to write about Brazilian culture and society on a sweeping scale. If we do not accept that there are some general patterns then we cannot even talk about Brazilian culture (or any other complex culture for that matter).

Distilling the principal cultural patterns of Brazilian society is a much more daunting task than generalizing about its history, regions, politics, or economy. I have chosen to focus on a select group of features that I feel are at the core of Brazilian society and culture: class, color, family, gender, sexuality, religion, carnival, music, television, film, literature, and soccer. These are features that define Brazil and Brazilians. All of them are simultaneously obvious and elusive. They are easy to identify but difficult to pin down. Having made this disclaimer, I generalize in the following pages while trying to provide a somewhat nuanced picture of what I believe are the defining characteristics of Brazilian society and culture.

RICH AND POOR

The starkest of these contrasts—and the one that has dominated Brazilian society from its beginnings in the sixteenth century—is the glaring disparity between rich and poor. In a country that has become one of the most industrialized in the world, and ranks among the ten largest economies in the world, roughly 30 percent of the population live in abject poverty, making less than $100 a month. Another 30 percent qualify as poor (by Brazilian standards), making less than $300 a month.

Although Brazil is the tenth largest economy in the world, it ranks seventy-fourth in per-capita income (about $2,700 in 1990). The historical divide between rich and poor has been growing in recent decades. According to a study by the World Bank, Brazil has the most inequitable income distribution of any country in the world. The richest 20 percent of the population earn twenty-six times as much as the poorest 20 percent of the nation. (The comparable figure for India is 5 to 1, and 11 to 1 for the United States.) Brazil is a rich country full of poor people.

In the United States, where sharp contrasts also divide rich and poor, nearly 70 percent of the population fall in the middle class, and about 20 percent at poverty or below the poverty line (about $1200 a month income for a family of four). Like the United States, Brazil has an upper class of some 10 percent of the population. Unlike in the United States, in Brazil the middle class accounts for just another 20 percent of the population. The middle and upper classes, then, form a minority of affluent citizens atop an immense mountain of poor Brazilians.

And a very big mountain it is indeed. Brazil entered the 1990s with a population of 150 million, making it the fifth most populous nation in the world (after China, India, the United States, and Indonesia). Brazil's population has doubled in the last thirty years and is nine times larger than it was at the beginning of the century. In the 1950s and 1960s, Brazil's population grew at the explosive rate of about 2.5 to 3 percent a year. A wide variety of factors, including the increasing use of contraception and family planning, brought the growth rate down to just 2 percent per year in the 1980s. According to some demographers, a rapid decline in the birth rate in the 1960s and 1970s amounts to a "baby bust." In the 1960s, the typical Brazilian woman in her childbearing years gave birth to six children. By the 1990s, the rate had dropped to three. This has helped lower population growth to a rate characteristic of the most industrialized nations (just under 2 percent). In just three decades, Brazil has made the so-called demographic transition (from high birth and death rates to low birth and death rates) that European societies took a century to complete prior to World War I.

Although the rate will most likely continue to decline in the coming years, Brazil will still see its population grow. Half the population is under the age of 20, and as this enormous "baby boom" generation moves through its childbearing years, population growth will be significant even if this large segment of the population does no more than reproduce itself. Rapid population growth over the past few decades has not caused the enormous poverty in Brazil. It has, however, severely aggravated a very old problem.

The "social question," as Brazilians call it, has been a problem since the early colonial period. This sharply divided social structure first emerged out of a plantation society built on a few landowners' dominating masses of non-white workers. Independence in the early nineteenth century did not dramatically alter this profound social division, but reinforced it as the expansion of coffee cultivation in the Southeast replicated the pattern of sugar in the Northeast. Although the appearance of mass communications, modern transport systems, and the rise of mass politics have eroded the power of the rural landowners, in many parts of the Brazilian interior the traditional power structure in the countryside remains intact.

In the backlands of the Northeast, the centuries-old tension between the subordinated majority (landless peasants) and the dominant minority (landowners) fuels violence. The traditional patronage networks of paternalistic landowners and the peasant laborers who depend on them for work, charity, and survival remain intact in many places. For centuries, the power of these landowners has rested on their ability to control access to land and the landless labor force, and to deny peasants access to the political system.

Historically, rural laborers with no access to land of their own have had no choice but to work for the large landowners. Despite the dramatic transformation of the economy, and the industrialization of major cities during the past fifty years, the structure of land ownership in the countryside has remained profoundly inequitable. According to government census figures, in the mid-1980s less than 1 percent of the farms (all more than 1,000 hectares, or roughly 500 acres) in Brazil accounted for more than 40 percent of all occupied farmland. Conversely, about 50 percent of all the farms (all smaller than 100 hectares, or roughly 50 acres) comprised 3 percent of occupied farmland. In short, a few large landowners control the bulk of occupied farmland.

In the past few decades, activists have repeatedly attempted to mobilize peasants to fight for access to land and to break down the unequal distribution of property. All major efforts at land reform have failed. Consequently, the rising demand for land and the increasing politicization of the peasantry, combined with the intransigence of large landowners, have led to an escalation of violence in the countryside. In the 1980s, more than 1,000 Indians and peasants died in clashes with the heavily armed landowners' henchmen. Only about a dozen men have ever been tried for these murders, and only half of those have served any time in prison. In one three-week span in August 1995, police and hired gunmen killed two dozen squatters who had staked out land claims in the state of Rondônia near the Bolivian border. Brazil's outmoded land structure continues to create social conflict in the 1990s, and the possibilities for land reform seem as remote as ever.

Even the rise of major urban centers during the past century has not substantially eased the problem. In the 1960s, Brazil became a predominantly urban society, and today 75 percent of all Brazilians live in urban areas. As millions of poor people abandon the countryside, the urban economy cannot absorb them. Unlike in Europe and the United States during the nineteenth century, industrialization in Brazil has not advanced rapidly enough to provide work for these rural migrants. Enormous shantytowns (*favelas*) with millions of unemployed and underemployed poor people are the end results.

Until the 1960s, successive governments attempted to fight the growth of city slums through traditional methods of "urban renewal." City planners tried to raze the *favelas* and relocate the poor in government housing, which was often miles from the inner city. By the 1970s, it had become clear that this process was not working and that the tidal wave of poor migrants had overwhelmed any pretense of relocation to public housing. In the 1970s and 1980s, the military regime tried to defuse the pressure on both the overcrowded rural areas and the exploding cities by promoting immigration to the vast Amazon basin. Over the past two decades, the strategy has shifted to efforts to provide the *favelas* with paved roads, electricity, sewage, potable water, schools, and clinics. Although much has been accomplished in "urbanizing" many of the *favelas*, the task sometimes seems overwhelming.

The Brazilian government long ago created a public health-care system similar to those of the industrialized nations. Those who have dealt with health care for the poor in the United States know how inadequate and overwhelmed the system is in its efforts to deal with 15 to 20 percent of the population. The public health-care system in Brazil is even more rudimentary—and it tries to cope with the needs of 60-65 percent of the population. The political lack of will to invest in basic health care creates high social costs in the long run. Brazil spends little on health care and spends it poorly. Brazilian government spending on health care in 1992 averaged about $50 per person per year. The United States, a country with a mediocre record for government spending on health care, spent fifty-six times more ($2,840) per person while having a gross domestic product that is only thirteen times larger than Brazil's. Argentina, with an economy less than half the size of Brazil's, spends about $300 per person per year on health care.

Despite enormous obstacles, in general Brazilians now live longer and healthier lives than their ancestors. Life expectancy has risen from just over 40 years in the 1930s to 66 years by 1990. Brazil, however, has entered into an epidemiological transition. Malaria, tuberculosis, and cholera—diseases typically associated with developing countries—remain major causes of death while cancer and heart disease—typical of developed economies—have also become major killers. Infant mortality rates have dropped dramatically, although they remain very high (50 per 1,000 live births) by the standards of the industrial nations of the Northern Hemisphere (about 9 per 1,000). Brazil also fares poorly when compared with other Latin American nations. Tiny Paraguay and El Salvador, countries with less than half the per-capita income of Brazil, both have lower infant mortality rates. About 1 million Brazilian children die each year before reaching their fifth birthday.

The benefits of improving health have been shared unequally by class and region. While those in the affluent classes can afford state-of-the-art facilities and services, the poor are lucky to receive basic immunizations and emergency care. Many Brazilians do not have access to even the most basic health care. Regional inequities stand out most sharply when one compares the Northeast with the Southeast. In 1980, fewer than one-third of all households in the Northeast had piped water, half the percentage in the Southeast. Only one in six houses were hooked up to sewers or septic tanks, while the figure was more than one in two for the Southeast. The infant mortality rate in the Northeast is more than double that of the

Southeast, and Northeasterners have had (through the 1980s) an average life expectancy seven years shorter than Southeasterners.

The geography of hunger further highlights the class and regional inequities. Thirty-two million—or one in five—Brazilians go hungry each day. Roughly half of those live in cities, and half in the countryside. With only a quarter of the population residing in rural areas, this means that roughly one of every two rural Brazilians suffers from daily hunger, while the rate for the urban areas is roughly one out of seven. Sixty percent of the hungry live in the Northeast. Brazil produces enough grains (rice, beans, wheat, corn, soy) to feed its population. The problem is not production, but inequitable distribution. With millions of starving people, Brazil has become one of the three largest exporters of agricultural products in the world.

Class inequities also reinforce a bleak pattern of educational inequities. Much like the U.S. South, Brazil has never placed a high priority on public education, nor has it invested intensively in public schools. The elites have always been able to afford schools and excellent instruction for their children. Educating the masses has never been a high priority. The first few public primary and secondary schools appeared in Brazil in the nineteenth century, and then principally served the children of the elite in the major cities. In the countryside, the affluent hired private tutors or sent their children to boarding schools in the major cities.

The same pattern followed with university education. Throughout the nineteenth century, Brazil had law, medicine, and engineering schools, but no modern liberal arts universities. Beginning with the creation of the University of São Paulo in the 1930s, the states and the federal government slowly created modern universities open to all—in theory. In reality, most Brazilians cannot afford to take the time to attend classes during the day. They must work. Public universities, then, have become the cheap educational route for the children of the affluent who graduate from expensive elementary and secondary schools. Ironically, the less affluent must pay relatively high fees to go to private (usually Catholic) universities that have tailored their courses and programs to students who cannot attend full-time during the day.

Government spending patterns reinforce the inequities of an already skewed educational system. Two-thirds of all public funds spent on education are funnelled into universities. Most university students come from families with incomes thirty times the minimum monthly salary, about $3,000 a month. The other third goes to primary and secondary education. The vast majority of students at the primary level come from families making less than the minimum salary of $100 a month. The government subsidizes the rich and throws crumbs to the poor.

The odds are heavily stacked against the less-affluent. Most cannot afford to wait for the long-deferred benefits of education. They have to survive in the present rather than waiting for the future. Most Brazilian children enter public schools, but few stay for very long. They quickly move into the labor force, usually into the informal sector selling goods on the streets or doing odd jobs. A recent study, for example, cited some astonishing statistics on Brazilian education. For every 100 students who entered the first grade in 1978, only 55 made it to the second grade, a dropout rate comparable to Mozambique and Haiti. Just 29 made it to the sixth grade, and only 6 of the original 100 entered the university eleven years later in 1989.

This abysmal record has hurt and will continue to handicap Brazil in the coming years. As the example of the East Asian countries has so vividly demonstrated in the past few decades, the most important investment a society can make is in the education of its people. Again, much like the U.S. South, Brazil is paying and will continue to pay for placing so little value on education. The failure to invest in human capital will be the most enduring and intractable legacy of underdevelopment.

The rural exodus and the growth of the *favelas* has transferred many of the tensions and pressures of the social question from the countryside to the cities. In the city of Rio de Janeiro, nearly 1 million of its 7 million inhabitants live in *favelas*. The city of São Paulo has nearly 600 *favelas,* and in Recife the *favelas* contain more than 40 percent of all the city's housing. Although many of the *favelas* have become permanent communities with paved streets, running water, and basic infrastructure, the newer *favelas* often spring up on rugged terrain, with makeshift housing, no running water or electricity, and few municipal services. The slums and shantytowns of the big cities have become breeding grounds of discontent, despair, and violence.

In the past decade, they have also become the focal point of the drug trade, especially in Rio de Janeiro. Powerful criminals have emerged out of the *favelas* to control the distribution of drugs, primarily marijuana and cocaine. The drug rings employ many *favelados,* especially children and teenagers, and bring large sums of money into the slums. Drug lords have become so powerful in the larger *favelas* that local police often either accept payoffs to leave them alone or simply refuse to go into the neighborhoods. (In February 1996, Michael Jackson and Spike Lee had to pay the drug lords to allow the filming of a music video in the Rio favela, Dona Marta.) In 1994, the federal government declared a national emergency and ordered army troops with armored vehicles into the *favelas* to combat the drug smugglers. Violent struggles among competing drug lords have also driven up the number of homicides and assaults.

Nova Iguaçu, a city of 1 million on the northern fringe of Rio, tallied 2,500 homicides in 1989, giving it the dubious distinction (according to UNESCO) as the most violent city on the planet. This converts to a rate of 250 murders per 100,000 population or more than three times the homicide rate of New Orleans or Washington, D.C., the cities with the highest murder rates in the United States (and surely the "murder capitals" of the industrial world). (As another point of comparison, metropolitan São Paulo tallied 20 homicides per day in 1995.) As in the United States, the greatest victims of violent crime in Brazil are the poor. Violence generated by the poverty and despair of life in teeming *favelas* has made the slums more dangerous than the most violent sections of the inner cities of the United States or Europe.

The economic crisis of the 1980s has also been accompanied by a rise in juvenile crime. Of the nearly 60 million children and adolescents in Brazil, half come from families making less than half the minimum wage (which generally runs about $100 a month). Another quarter live in families making less than the minimum wage. In other words, as many as 45 *million* children live in what the Brazilian government defines as extreme poverty. Possibly as many as 10 million children now live on the streets of Brazil's major cities, and many have resorted to crime to survive.

Hired "death squads" that became notorious for acting as vigilante groups in the 1970s have begun to target adolescents and teenagers for assassination. Merchants and shopkeepers who see the youths as a threat to their business, property, or own personal safety sometimes hire gunmen (often off-duty or former policemen) to "eliminate" the "troublesome" adolescents. These vigilantes act as judge, jury, and executioner. According to estimates by Amnesty International, three or four children are murdered on the streets of Brazilian cities each day by the so-called justicers *(justiceiros)*. The majority of those killed by vigilantes are non-white teenagers between the ages of fifteen and seventeen. The troubles of Brazil's children, this most precious of resources for any society, perhaps foreshadow the enormous difficulties facing Brazilians in the coming years.

The appearance of death squads is a symptom of the much larger problem of a lack of public confidence in the police and criminal justice system. Historically, the police (and other security forces) have been a crucial element in maintaining elite control over the masses. They have not been reticent in using force, especially against the lower classes. Torture and beating lower-class suspects has long been routine, while wealthy criminal suspects rarely see the inside of a police station. The affluent and the police often act with impunity. In 1992, police officers shot and killed 1350 people in the city of São Paulo alone. Few of the shootings are ever investigated. Most of the victims were poor.

Brazilians are well aware of the class inequities in the enforcement of laws and the treatment of suspects by the police. In recent years, public confidence in the criminal justice system has perilously eroded in the face of rising crime in the major cities, and the inability or unwillingness of the police and courts to confront the problem. According to the Institute of Religious Studies (ISER) in Rio de Janeiro, more than 90 percent of all homicides in the city go unpunished. This does little to build public confidence in the police and courts. Increasingly, the "people" *(povo)* have begun to resort to vigilante justice as with the death squads. Lynchings have also been on the rise in the last decade. As drug trafficking, homicide, and robbery have become more common, outraged citizens, especially in the *favelas,* more and more frequently take justice into their own hands.

In contrast to the chaotic and violent drug trade and life in the *favelas,* the illicit "numbers" game that pervades urban life has become highly popular, and all but accepted by most Brazilians. This clandestine lottery operates across the country, with its major centers in the cities of São Paulo and Rio de Janeiro. The game began in Rio at the turn of the century, when the Baron João Batista Drummond took the lottery idea from a Spaniard to raise money for his private zoo. The baron substituted the flowers on the Spaniard's lottery tickets with animal figures. Ever since, the masses have bet on numbers and animals in what came to be known as the *jogo do bicho* ("animal game").

On many occasions during this century, the police have unsuccessfully attempted to stamp out this form of gambling. In the late 1940s, the Brazilian government went after the game with a vengeance. Yet, despite the repression, the popularity of the game has increased over the years. The 1980s were a golden age for the game. The government all but declared the game legal, and business and profits boomed. Recent estimates place the volume of proceeds from betting at $150 million a month, and Rio de Janeiro alone has more than 300 "bankers" employing more than 45,000 people. Profit margins are estimated to run at somewhere between 15 and 30 percent. The so-called animal bankers (*bicheiros*) who control the lottery have become wealthy and (at times) accepted figures in elite society. The names and territories of the major bankers appear regularly in the society columns of newspapers, magazines, and television. In the 1990s, the police and courts have once again cracked down on the *bicheiros*, partly due to their growing involvement in drug trafficking.

The game employs thousands of people in the major cities. On virtually any major street in Rio de Janeiro, one can see people placing their bets with the *anotadores* ("note takers"), who then pass them on through a series of individuals and runners. (The minimum bet is about 25 cents.) At precisely 2:23 every afternoon, at a location that changes daily, a group of witnesses and *bicheiros* gather to spin a globe with ten balls numbered zero to nine. As each number drops from the globe, the results are relayed by phone across cities and states. An estimated 10-15 percent of *cariocas* and *paulistanos* play the game on a weekly basis. The integrity of the game has become so well established that in one recent opinion poll Brazilians gave the *jogo do bicho* a higher confidence rating than the postal system, the phone system, the Catholic Church, and the other major institutions in the country. Like lotteries elsewhere, news of the big winners travels fast and fuels betting. Unfortunately, the *jogo do bicho*, like all lotteries, offers an illusory dream of wealth for all except a small number of the bettors.

The *favelados* who do not win the lottery and who turn away from the allure of "easy" money in the drug trade face few alternatives. The cycle of poverty so familiar to North Americans traps the Brazilian poor with an even greater vengeance. Schooling is limited and children rarely attend for more than a few years. These poorly educated, unskilled laborers then compete with millions like themselves for scarce jobs. Teenage pregnancy is on the rise, as are female-headed households and demands on the limited social services of the municipal, state, and federal governments. Facing few opportunities in life, many turn to illicit activities to survive.

This scenario has dire implications for social and political life in a developed nation where 20 percent of the population is poor or near poverty. In a developing nation where the figure is 60 percent, the consequences are disastrous and tragic. Think of how the United States and European nations grapple with the social and political costs in societies where the poor are a minority and then imagine the numbers of poor and affluent being reversed. In the United States, the majority struggles to find ways to incorporate a 15-20 percent minority into the mainstream. In Brazil, the poor are the mainstream, and the 20 percent minority has shown little interest in confronting the enormous challenge of incorporating the majority into the benefits of modern, urban society.

Despite decades of impressive economic growth, the striking social inequities remain. In a recent survey of 1,500 of the most influential members of Brazil's political and economic elite, close to 90 percent believed that Brazil had achieved economic success and social failure. Close to half viewed the enormous inequities as a form of "social apartheid." Perhaps most ominous, two-thirds of this elite group believed that in the 1990s the country would experience a state of chronic social convulsion. Brazil pays a very high price for its profound social divisions.

A SPECTRUM OF COLORS

Throughout the Americas, racial discrimination has reinforced the inequities of class in regions that imported large numbers of African slaves: Brazil, the Caribbean, and the United States. As in the Caribbean, three centuries of slavery and the slave trade profoundly Africanized Brazilian culture and the nation's population. Slave traders shipped nearly 4 million Africans across the Atlantic to Brazil from the mid-sixteenth century to the mid-nineteenth century. Brazil became the final destination of more than one-third of all the African slaves brought to the Americas. (The figures for the Caribbean are slightly higher. In contrast, recent estimates place the number of African slaves brought to the United States at around 750,000, or around 6 percent of the total slave traffic.)

The constant flow of Africans into a colonial society with a small European population created a black majority in Brazil by the early 1600s. The majority of the Portuguese who crossed the Atlantic were single males. The scarcity of white women, and the power of white men over African slaves, rapidly resulted in a racially mixed population. Very quickly a broad color spectrum developed that ranged from black to white with complex gradations in between. The intermixing of Indians and Portuguese, and Indians and Africans, added to the complexity.

The offspring of blacks and whites—the mulattos—became a substantial proportion of the population by the late colonial period. At the beginning of the nineteenth century, half of Brazil's population of 3 million were enslaved blacks, one-quarter were whites, and one-quarter free blacks and mulattos. The slave trade continued to accelerate until its abolition in the 1850s, thus reinforcing the "black" majority. According to official censuses, the white population did not rise above 50 percent until the beginning of the twentieth century. The most recent census (1991) places the number of blacks at 5 percent and mulattos at 39 percent. (Asians, under the color category "yellow," comprise about 1 percent of the population.) Although "whites" have now become the majority (55 percent), one should keep in mind how color categories are culturally determined in censuses. To be black in Brazil means to have no white ancestors. Many "whites" have some racially mixed heritage. Further complicating matters, the color of census respondents is "self-described," with each respondent deciding his or her own category. (One recent social science survey came up with more than 200 race and color terms in use.)

While racial mixture and the rise of a mulatto population also took shape in the Caribbean, the United States moved toward a more inflexible system. To be considered white in the United States, one cannot have any non-white ancestors, and many North American "blacks" have some racially mixed heritage. Although the British in North America did not refrain from sexual relations with blacks, Anglo-Americans rarely publicly acknowledged their racially mixed offspring.

In the United States, a rigid, bipolar racial division took shape. North Americans distinguished between blacks and whites, allowing little middle ground. Being black was also virtually synonymous with being a slave. On the eve of the Civil War, a mere 6 percent of the black population in the United States was free. At the same time in Brazil, the majority of the black and mulatto population lived in freedom.

Race relations in Brazil and the United States have diverged largely due to this striking difference. Before the Civil War, slavery divided blacks and whites in the United States. With emancipation, whites turned to Jim Crow laws and segregation to separate and discriminate against blacks. For blacks in North America, the path from slavery to freedom became the path from slavery to racial segregation. The struggle for freedom became the fight for political rights and inclusion in the system.

The Brazilian experience demonstrates that other paths were possible. For centuries, racial mixture created a complex color spectrum and a large, free black and mulatto population. These free blacks and mulattos lived and worked alongside whites. Despite color consciousness and discrimination, Brazilians forged a racially mixed society instead of a racially divided one. Brazilians discriminated, but on the basis of color, and there were many shades. North Americans discriminated on the basis of race, and there were but two. With the emancipation of the slaves in Brazil in 1888, this complex racial mixture and color spectrum made the construction of legal segregation impractical, if not impossible. How could the separation be defined, much less maintained?

During the past century, as the United States grappled with segregation and then painfully battled over its dismantling, many of the Brazilian elite watched the struggle in the United States with a sense of self-satisfaction with both their own "racial harmony" and the lack of state-supported racial discrimination. This is not to say that racism does not and has not plagued Brazil, but it is a racism that has been supported by informal collusion rather than by law. As one well-known Brazilian saying puts it, "we don't have a racial problem here, because blacks here know their place."

As in the United States, the political and cultural elite of nineteenth-century Brazil discussed the "race question," and the racism of the period was both blatant and widespread. Many abolitionists in the late nineteenth century desperately wanted to end slavery and the slave trade so that Brazil could halt the influx of Africans and begin to "whiten" the population both biologically and culturally using European immigrants. They believed that their country could never be "modern" as long as a large percentage of the population had non-European origins.

At the turn of the century, the consequences of Brazil's African heritage obsessed intellectuals, and most despaired over its powerful impact. Most believed racial mixture had condemned their society to backwardness, if not utter hopelessness. In the 1920s, the work of Gilberto Freyre began to fundamentally reshape the debate over race and the national psyche. In his many books and articles, Freyre transformed the debate over race. The African influence in Brazil had not held the nation back, he declared, it had made it stronger. Without the African mixture, the Portuguese would have neither survived nor thrived in the tropics. The Portuguese created a "new world in the tropics," and the mixture of African, Indian, and European made Brazil unique. According to Freyre, this "lusotropical" civilization had produced a racial democracy in which blacks, whites, Indians, and the racially mixed mingled harmoniously. (Educated at Baylor and Columbia Universities, Freyre had witnessed firsthand race relations in the United States, and he consciously drew the distinction between the two countries.) While privately racial prejudice did not disappear, publicly Freyre's view gradually became the official vision and a basic component of the national psyche. As the civil rights movement in the United States tore at the social fabric of the nation, Brazilians (especially the elites) congratulated themselves on what they perceived as the racial harmony and lack of segregation in their society.

Racial prejudice and discrimination exist in Brazil, but historically they have been more subtle and complex than in the United States. Many employment advertisements in newspapers speak of "good appearance," a widely understood code phrase for light skin. Surveys regularly point to strong attitudes about color consciousness that are reinforced by social class. To oversimply a complex issue, in Brazil racial prejudice seems to intensify as one moves up the social scale. In the United States, the opposite seems to be true. In Brazil, the elite's racial prejudice toward those of darker skin, however, has been moderated by economics. The upper classes in Brazil have historically been less reluctant to accept non-whites into their ranks when wealth produces social mobility. As the famous Brazilian saying goes, "Money whitens."

Census figures quickly dispel any doubts about the existence of some form of racial prejudice and discrimination. In a society in which nearly half the population is Afro-Brazilian, most of that half lives in poverty. The pattern has been clear for centuries. Most of the poor in Brazil are dark, and most of the affluent are white. Nearly 30 percent of all blacks and racially mixed persons are illiterate, while the rate for whites is 12 percent. Whites have an average monthly income nearly two-and-one-half times that of blacks, and twice that of mulattos. (In the United States, family income for whites is nearly double that for blacks.) The statistics are too striking to be coincidental.

Public life reinforces the impersonal census data. Even more so than in the United States, most successful blacks can be found in the arts, entertainment, and sports. Dark-skinned diplomats and politicians are a rarity, even more so now than in the nineteenth century. Nonwhite, especially black, actors and actresses have been a rarity on prime-time evening television. In 1990, just 7 congressmen (out of 559) considered themselves black, and only 6 of Brazil's 362 bishops were black in a nation with a population of 70 million blacks and mulattos. Out of nearly 13,000 Catholic priests, only about 200 are black.

A small but persistent black consciousness movement took shape in the decades following abolition and has reemerged in the past two decades, but it has failed to attract widespread support. Here is the pernicious side of the myth of racial democracy and the widespread racial mixture. How is it possible to build a movement around consciousness of being black when most non-whites do not see themselves as black and do not wish to be considered black? After all, whiter skin is widely believed to be a key to mobility. Mixture and assimilation have made it difficult, if not impossible, to build a movement based on a sense of separateness.

In many ways, race relations in Brazil and the United States have been converging during the past two decades. Recent studies seem to show that the so-called three-tier system of race relations in Brazil (blacks, mulattos, whites) has increasingly become more of a two-tier system. On the one side, whites tend to view blacks and mulattos as a single group of Afro-Brazilians. On the other side, blacks and mulattos seem increasingly to view themselves as a group apart from whites, despite the enormous variations they see among themselves.

Conversely, the traditional solidarity of blacks in the United States has been strained by the successes of the civil rights movement. Now that blacks in the United States have achieved legal and civil rights, and as increasingly larger numbers achieve economic success, what will be the basis of a common movement? Will a sense of racial identity be enough? Will this identity continue as the forces of integration and assimilation increase? Although the forces for solidarity remain weak within the Afro-Brazilian population, and the strains on black solidarity in the United States have just begun to appear, race relations in the two largest American societies share greater similarities now than perhaps at any time in their histories.

Throughout the twentieth century, both blacks and whites in Brazil have feared that their numbers were diminishing and that their racial groups were in danger of being overwhelmed by the racially mixed. Up until 1870, Brazil was overwhelmingly a "black" nation. A massive wave of more than 2.5 million European immigrants—mainly from Italy, Portugal, and Spain—shifted the racial balance in the late nineteenth century. A large number of Eastern Europeans, principally Jews, also entered Brazil during these years. Middle Eastern (mainly Lebanese) and Japanese immigrants added to the rising tide of immigrants by the beginning of the century. Many intellectuals welcomed this shift and believed that by the end of the twentieth century immigration would "whiten" the population and gradually eliminate Brazil's black heritage.

By 1950, census takers counted only 11 percent of Brazilians as black (preto), while they classified 62 percent as white (branco). Both categories had diminished by 5-6 percent by the end of the 1990s, as the racially mixed (pardos, largely mulatto and caboclo) population increased to 39 percent of all Brazilians. An ever-increasing number of those in the racially mixed category are describing themselves as brown (moreno) rather than mulatto. Ironically, as the country seems to be moving toward a clearer consciousness of the division between white and non-white, fewer and fewer Brazilians consider themselves black, and more than one-third are neither black nor white.

Brazil, much more so than the United States, is a racial "melting pot." Brazil has become a racial rainbow, a continuum of shades and colors. Yet, racial and cultural mixture have not eliminated racial prejudice and discrimination. The overt racism of the elites prior to the twentieth century clearly negates Gilberto Freyre's vision of Brazil as a racial democracy, as does the more subtle racism of the contemporary era. Despite its flaws, however, Brazil has forged one of the most racially mixed societies in the world.

Indians and Asians (together accounting for less than 1 percent of the population) are the only racial groups in Brazil that experience a type of segregation and polarity similar to that in U.S. society. They are also the oldest, and newest, immigrants to the land. Until very recently, the Asian community has been overwhelmingly made up of the Japanese, who began to arrive in Brazil in 1908, principally to work on the plantations of the São Paulo interior. Between the two world wars, a quarter of a million Japanese immigrated to Brazil, and today the Japanese community is the largest in the world outside of Japan. About one million Brazilians have Japanese ancestry, and 80 percent of them live in the Southeast. Nearly 70 percent live in the state of São Paulo. Although they came to Brazil to work in the fields, most Japanese-Brazilians today live in urban areas, and São Paulo has a "Japan Town" that dwarfs similar communities in San Francisco, Los Angeles, and New York City.

Like their North American counterparts, Japanese-Brazilians have become exceptionally successful. Second- *(nisei)* and third-generation *(sansei)* Japanese-Brazilians have become cabinet ministers, major bankers, and renowned artists. Asians are the single most successful ethnic group in the country, with an average monthly income nearly twice that of whites. As in the United States, the Asian community has worked hard to maintain its ethnic and cultural identity through schooling and intermarriage. More than 60 percent of the Japanese-Brazilian community speak Japanese, and another 10 percent claim to understand the language. As in the United States, however, the slackening flow of immigrants in the past forty years and the forces of assimilation have begun to make it more difficult to retain language and marriage patterns.

The native peoples of Brazil have also begun to feel the pressures of assimilation more intensely than at any period since the conquest in the sixteenth century. Estimates of the size of the indigenous population in 1500 range from 1 to 5 million, and they are nothing more than guesses given the lack of historical evidence. The "Indian" population of Brazil today is about 250,000, with two-thirds of it living in the Amazon basin. Much like Indians in the United States, Brazilian Indians had not achieved the complex stages of political organization found in Mesoamerica and in the Andes, where the Aztecs, Mayas, and Incas built sophisticated empires around dense populations in the millions. As in North America, the Indian populations were highly dispersed and mostly semi-sedentary or nomadic tribes. (An estimated 5 million Indians lived in the area of the present-day United States in 1500. Some 1.3 million Indians, most of whom are very acculturated to non-Indian society, live in the United States today.) From the arrival of the Europeans until the twentieth century, European diseases and warfare devastated Indian tribes, and governments forced the survivors onto reservations.

As in the United States, these reservations were located in areas of little interest to whites. In Brazil, this meant the seemingly impenetrable forests of the Amazon. At the beginning of this century, Colonel Cândido Rondon of the Brazilian army became a national legend as he crisscrossed the interior in an attempt to establish contact with the remaining tribes. Rondon formed the Indian Protection Service (SPI), which exercised a paternalistic control over the tribes, many of whom had retained stone-age cultures in complete isolation from the modern world.

The SPI was transformed into the National Indian Foundation (FUNAI) by the military regime in the 1960s and became infamous for its abuses of power. The military regime saw the Amazon basin as the key to the future of the country, and it set out on an ambitious program to develop the resources of the region. As colonization programs and multinational corporations moved into the region, and roads were built, the Indians came under enormous pressure. Hydroelectric projects threatened to flood traditional Indian lands; prospectors (*garimpeiros*) in search of gold and peasants in search of land poured onto the once-isolated reservations. In 1989, on the northern frontier with Venezuela, nearly 20,000 *garimpeiros* flooded into the territory of Roraima and invaded lands of the 10,000-member Yanomami tribe. The results have been devastating. Anthropologists expect this stone-age tribe to disappear within the next decade if drastic protective measures are not taken soon.

Indians and their allies have been fighting back through the Brazilian legal system and through direct action. Several Indians have been elected to state-level offices, and Mario Juruna, a Xavante chief, was elected a federal congressman for the Democratic Workers Party in 1983. A colorful figure who speaks heavily accented Portuguese, Juruna sometimes appears in public in traditional Indian dress and body paint. The Kayapó tribe in the south of Pará state has been battling an enormous hydroelectric project for years. Their struggle has gained international attention and angered those who see the development of the Amazon basin as an inevitable and essential process for the nation's future.

Brazil's dwindling indigenous population faces a reduced set of choices: acculturation, isolation, or some sort of mediation between the two. Given the enormous reach of modern communications technology and given the pressures by those favoring development, the option of isolation has disappeared. The best the Indians can hope for is to strike a bargain with outsiders to maintain as much of their traditions as possible, and this hinges on their ability to retain land rights. The pressures are enormous. Reservations (at least on paper) cover 200 million acres, or nearly 10 percent of the nation's territory. The Kayapó alone, a tribe of some 1,700, have title to more than 7.5 million acres. The landless peasants of the Northeast and the "modernizers" of the big cities see this as an enormous imbalance and a waste of national resources. It would seem that the Indians of Brazil will most likely experience a process of acculturation similar to that of the Native Americans to the north. They may keep their reservations and many of their traditions, but it is hard to imagine the survival of stone-age cultures living in splendid isolation. Five centuries after the arrival of the Europeans, the peoples that once occupied all of Brazil now number less than two-tenths of 1 percent of the Brazilian population.

VARIETIES OF RELIGIOUS EXPERIENCE

Brazil's own particular Catholic heritage has played a large role in shaping the national culture. Like the Spanish, the Portuguese imposed Roman Catholicism on the Native Americans and Africans they enslaved. Throughout more than three centuries of Portuguese imperial rule, Catholicism reigned as the official religion of the colony, and crown and clergy periodically attempted to stamp out traces of other religions and religious practices. The Portuguese and Spanish monarchs had a special arrangement with the Vatican that allowed them to control the appointment of high church officials in their domains. This "royal patronage" (padroado real) gave the crown control over the appointment of church officials in the Portuguese empire, and the church became, in effect, an arm of the state.

Despite independence from Portugal in 1822, the system of royal patronage effectively continued under Pedro I and II. Although Roman Catholicism remained the official state religion, the constitution of the Brazilian Empire (1824) established religious freedom. Protestant religious groups gained a foothold in the country, although they primarily served the foreign community, especially the British. During the Second Reign (1840-89), Pedro II kept tight control over the Catholic Church. Although the official head of the Catholic Church in Brazil, Pedro was privately an agnostic and wished to restrain the political and cultural influence of the Church as much as possible. Pedro squeezed the Church, severely limiting the recruitment of new priests and refusing to allow the creation of new dioceses. The government collected tithes for the church, and then paid church officials. In effect, they became civil servants under government control.

This close control and regulation diminished the institutional presence of a Roman Catholic Church that had never been very powerful. Unlike in Mexico or Peru, where the Church had become a very powerful economic and political institution during the colonial period, in Brazil Portuguese monarchs had never allowed the Catholic Church to build up much power. Consequently, Pedro II weakened an already frail institution. By the end of the Empire (1889), the Catholic Church had just 12 dioceses to serve a population of about 12 million. For every priest there were about 15,000 faithful (compared with 1 to 900 in the United States at the same time). By the end of the nineteenth century, the Catholic Church had virtually no political or economic power in Brazil.

Perhaps this political weakness helps explain the relatively tolerant approach of the Brazilian Church to other religions when compared with places such as Mexico. The seven-hundred-year religious and military struggle to drive the Islamic Moors from Spain forged the most aggressive and militant Church in Catholic Christendom by 1492. The Spanish carried this militant Catholic ethos across the Atlantic in the conquest of the Caribbean, Mexico, and Central and South America. The Inquisition reached Spanish America and pursued suspected heresy and heretics with much greater determination than the missionaries in Brazil. Having completed the reconquest of Portugal in the mid-thirteenth century, Portuguese Catholics did not always show the militant zeal of their Spanish counterparts. This less-militant form of Catholicism, coupled with a politically weak institutional presence, appears to have made the Portuguese empire more tolerant of religious deviation. Jewish merchants and Jewish converts to Christianity, for example, moved through the Portuguese empire with much greater ease and with less persecution than in the Spanish colonies. After independence, the Brazilian Church took a relatively tolerant attitude toward Jews, Afro-Brazilian religious practices, and Eastern religions such as Buddhism and Shintoism.

Although it had little political and economic power, the Church did have enormous cultural influence. By the nineteenth century, 95 percent of Brazilians were, at least nominally, Catholic. (With few exceptions, those who were not Catholic were mostly foreigners or Native Americans.) The Iberian traditions of Catholicism had been transferred to the New World and passed on for centuries to succeeding generations of Brazilians. Traditional Catholic values permeated and helped define Brazilian culture.

These values continue to define Brazilian culture today even when divorced from their Catholic origins. The strong emphasis on the male-dominated family, on procreation, and on community has its origins in Catholic traditions and theology. In contrast to the radically individual-istic Protestant traditions of Anglo America, Iberian and Iberoamerican Catholicism draw on collectivist and communal traditions deeply rooted in the medieval European church. North Americans tend to view society as a machine constructed of individual components. While each operates on its own, somehow the components come together to make the machine function smoothly. Iberian Catholicism stresses the organic nature of society and the need for collective solidarity. Society is a body (*corpore*) that cannot remain healthy and functioning unless its compo-nent parts work in harmony. These components are not all equal as individuals, but rather they all have their assigned place and function in a social hierarchy.

The anthropologist Roberto DaMatta has astutely pointed out that the egalitarianism of U.S. society and the hierarchy so characteristic of Brazil can be summed up in two simple phrases. For the United States: "Who do you think you are?" For Brazil: "Don't you know who you're talking to?" Brazilians live in a world of two superimposed systems of relations—one personal, one legal. Power comes from a strong hierarchical network of personal relations. Those without these connections are left to defend themselves through the impersonal forces of the legal system. Ironically, the law keeps the weak in their place, and the powerful use their connections to remain above the law.

In Brazil, Catholicism's greatest legacy has been to pass on these values of collectivism, hierarchy, male domination, family, and community. These values have shaped Brazilian culture and politics since the colonial period, and today they remain vital to understanding the national psyche, politics, and culture. With 85 to 90 percent of its population of 150 million professing at least nominal allegiance to the faith, Brazil has become the largest Catholic country in the world. Catholicism in Brazil, however, has been shaped by African and native cultures to produce a unique and complex variety of New World Catholicism. The historically weak institutional presence of the church, and the forced migration of millions of Africans to Brazil over three centuries, produced an interpenetration of religions as European Catholicism mingled with African and Amerindian religions.

Over centuries and across the vast regions of Brazil, Native American, African, and European religions competed, mixed, and developed alongside each other to produce a complex set of religious traditions. As the Europeans fought to impose their cultural and religious traditions on the conquered, Indians and Africans fought back, striving to retain their own heritage. The result is an extraordinary religious diversity in what is (at least nominally) an overwhelmingly Catholic nation. Today, tens of thousands of Indians on reservations continue to preserve some of their pre-conquest religious practices. In the heavily black areas of the Northeast, especially around Bahia, African religious practices survive, albeit in modified form. The coexistence of Native American, African, and European religions has also been accompanied by the emergence of uniquely Brazilian religions that draw on all three cultures.

The most visible and striking non-European religious practices are the survivals of African religious rites. With the exception of skin color, these religious rites stand out as the most visible reminder of Brazil's African heritage. The Afro-Brazilian religions testify to the enormous resistance of African slaves to efforts to strip them of their heritage and humanity. For three centuries, the Portuguese forcibly transported

Africans into slavery in Brazil and attempted to eradicate their culture and humanity. These black men and women and their descendants gradually adopted a new language, a new dress, and a new homeland, but they managed to retain elements of their cultural traditions, most prominently their religion. Forced to accept the Catholicism of the masters, many of the slaves secretly continued to practice their African religions. Brazil, in the words of the French sociologist Roger Bastide, became a land with a "white mask over a black face."

In a colony that was largely black and African, and that had few priests, generations of slaves and free blacks passed along African religious traditions. In some places (especially the Northeast), the African practices remained fairly pure and uncorrupted by outside influences. In other regions, African, Indian, and European religions mixed and formed blended, or syncretic, religions that were not African, nor Indian, nor European. *Candomblé* in northeastern Brazil is the most African of these religious survivals. According to Bastide, *candomblé* "is a genuine bit of Africa transplanted" to the Americas. In some areas of the northeastern and northern backlands, *candomblé* has mixed with Indian religions and is often called *candomblé do caboclo* (*candomblé* of the mestizo) to distinguish it from the more African version of the coast. *Macumba,* which is generally associated with Rio de Janeiro, offers another form of African religion that has arisen during the last century, one heavily influenced by European spiritism. The convergence of *macumba* and spiritism has forged a new Brazilian religion during the last century known as *umbanda.*

All of these religions draw heavily on African roots. Many of the Africans brought to Brazil came from the regions that roughly correspond to the modern nations of Nigeria and Angola. The Yoruba and Dahoman peoples of West Africa, and the Bantu of Angola, brought with them their polytheistic religions. These peoples venerated their ancestors and emphasized the worship of gods who controlled the forces of nature, and who protected their ancestors and homelands. They viewed life as a cyclical, never-ending struggle between the forces of good and evil.

Uprooted from homeland and kin, the Africans in Brazil could no longer retain the strong links between their gods and their ancestors. They continued to worship the gods known as *orixás* (oh-ree-shahs) in Yoruba. Africans probably worshipped some 400-600 of these *orixás.*

About 50 survived the transatlantic cultural passage. *Candomblé* today centers on some 16 *orixás,* and *umbanda* on 8. In Africa, Olorum was the supreme god who created the other *orixás.* (Note the similarity to classical mythology.) For example, he created the goddesses Iemanjá (to rule over the seas) and Iansan (to control the winds).

Many Africans, and some white priests, attempted consciously to bridge the divide between Christianity and African religions by emphasizing the similarities between the *orixás* and the saints of the Catholic faith. Thus, Iemanjá blended with the Virgin Mary, and Iansan with St. Barbara. Oxalá, the god of creation and procreation, was equated with Jesus, and Exu, who is usually identified with the forces of evil, took on the trappings of Satan. Although the blending of Catholic saints and African deities took place unevenly and not at all uniformly across Brazil, the process produced a powerful mix of African and European religions.

Both *candomblé* and *macumba* coexist with traditional Catholicism. Many of those faithful to these Afro-Brazilian religions also consider themselves good Catholics and see no contradiction between the different belief systems. In fact, many do not distinguish between the systems, seeing them as one. In one of Brazil's most famous plays (later an internationally acclaimed film), *O pagador de promessas* (literally, "the payer of promises," but translated as *The Given Word*), the central protagonist, Zé, illustrates the complete blending of African and European religions. He makes a vow to St. Barbara/Iansan that he will carry a cross on his shoulders to the altar of a cathedral in Salvador if she saves the life of his ailing burro. The animal recovers and he makes his pilgrimage to pay his promise. After a long and painful journey, he reaches the door of the cathedral only to be denied entry by the local priest, who denounces his promise as one made to a pagan goddess. Unable to see the distinction between Iansan and St. Barbara, Zé is baffled by the priest's stance. The conflict between the priest (representing the traditional church hierarchy) and Zé (supported by the lower classes) symbolizes the tension between an official, European religion of the upper classes and an unofficial, Afro-Brazilian Catholicism practiced by the masses.

If the beliefs of the two systems intermingle in the minds of many, the rituals are more clearly separated. *Candomblé, macumba, umbanda* and the like have their own houses of worship and their own hierarchies.

Candomblé ceremonies take place in the *terreiro* and are supervised by a spiritual leader, a *babalorixá* (if male) or a *yalorixá* (if female). They are the central religious figures who have the special powers to invoke the gods for the faithful. They have achieved their status not through seminaries or schooling, but through apprenticeship and the faithful community's recognition that they have special powers. Unlike Catholic clergy, they are not full-time employees of a church. Most have their own career or profession and handle their religious duties on the side. The *babalorixás/yalorixás* preside over "nations" within *candomblé* that once corresponded to specific African peoples. Membership in *candomblé* is very localized and specific rather than being universal as in Christianity or Judaism. One is initiated into a specific nation in a particular locale.

The basic ceremony of *candomblé* and *macumba* derives from the Nagô (Nigerian) nation. The faithful sacrifice animals to the god of the day, and then the spiritual leader invokes the deities in a fixed order. As they sing and dance, some people experience ecstatic seizures during which the gods take possession of their bodies and speak through them. Sometimes the god speaks intelligibly through the believer, and other times the language is unintelligible (similar to the "speaking in tongues" of North American Pentecostals). After people dance to the gods, the possession ends, and the worshippers complete the ceremony with a communal meal.

The *terreiro* ceremony has been intensely studied by Brazilian, European, and North American social scientists. It is practiced in its purest African form in the *candomblé* of the Northeast. In the past few decades, it has become a regular tourist attraction in the *macumba terreiros* of Rio de Janeiro. In the view of Bastide, "The Rio *macumba* is becoming more and more debased losing all its religious character and deteriorating into a stage show or mere black magic."

The *macumba* of Rio has also been transformed by the influence of European spiritism, especially the variety espoused by the Frenchman, Allan Kardec (the pseudonym of Léon Hippolyte Dénizart Rivail, 1803-69). Already a diluted version of *candomblé, macumba* blended with spiritism and its emphasis on reincarnation. The mixture of spiritism and *macumba* produced *umbanda,* which has grown rapidly, especially in the major cities of the Southeast during the past half-century. (Kardec's works continue to be influential and widely available.) Rather than serve African deities, *umbandistas* experience possession and can become channelers for many different spirits. The extremely emotional and violent nature of ecstatic possessions characterizes *umbanda.*

This brief and oversimplified survey of the blending and parallel development of Catholicism, spiritism, and non-Western religions barely begins to address the complex and unique melding of religions in Brazil. The interpenetration of religions has produced a culture that is at once Western and non-Western, that breaks down the barriers between black and white. While nominally Catholic and European, Brazilian culture has been profoundly influenced by African religious beliefs and values. The Afro-Brazilian religions (like African-American religion in the United States) are religions of joy and celebration that contrast with emphasis on suffering and asceticism in Anglo-American Protestantism and Spanish American Catholicism. Unlike the linear, historical vision of European Christianity, Afro-Brazilian religion offers a vision of a cyclical universe in which the struggle between good and evil never ends. The mystical and magical pervade Brazilian religious culture.

These values have profoundly influenced Brazilian culture and shaped the Brazilian psyche. Any Latin American specialist who has lived and worked in both Spanish America and Brazil becomes acutely conscious of the vast cultural divide between the two "halves" of Latin America. In broad terms, Brazilians are more easy-going, more optimistic, and less intransigent than their Spanish American neighbors, and I believe this difference has its roots in Brazil's blending of African, American, and Portuguese cultures.

This religious blending has also helped break down the distinctions between black and white. *Candomblé* carries on the traditions of Africa, and its adherents remain predominantly black and mulatto. Increasingly over the last century, however, the influence of African religion has permeated Catholicism, and through *umbanda* has passed into white society. Most *umbandistas* are light-skinned by Brazilians standards. *Umbanda,* in a sense, represents the transformation of African practices into a predominantly "white" religion. Yet, not all blacks practice African religions, nor do all whites adhere to traditionally European faiths.

Religion reinforces the dissociation between culture and race in Brazil. Where else but Brazil could one find whites accepting traditionally African religions and blacks rejecting them in favor of traditionally European faiths? As Bastide so aptly put it, "it is possible in Brazil to be a Negro without being African and, contrariwise, to be both white and African." For three centuries, black gods hid behind white masks. Today the masks have been stripped away and these black gods have been embraced by whites as well as blacks. Africa has been thoroughly integrated into the mainstream of Brazilian culture.

Over the last few decades, the Catholic Church has reached an unspoken accommodation with Afro-Brazilian religions. The two have achieved a form of symbiosis, and the old antagonism has virtually disappeared. During the last two decades, a new challenge has arisen for both Catholics and spiritists: evangelical and pentecostal Protestantism. Like the theology of liberation, this strain of Protestantism challenges the traditional religious dominance of the Catholic Church. Unlike it, the movement is politically conservative, staunchly capitalist, and supportive of the status quo. Conservatives, liberals, and radicals within the Catholic Church see the rise of this aggressive form of Protestantism as a threat.

The explosive emergence of Protestantism during the last twenty years has taken place in many Latin American countries—especially in Central America, Chile, and Brazil. Protestant missionaries, led by the Baptists, Methodists, and Presbyterians, began systematic work in Latin America over a century ago. In this overwhelmingly Catholic region, they made little headway. In 1930, only about 2 percent of the 100 million Latin Americans belonged to Protestant faiths. As late as 1960, the percentage had risen to around only 2.5 percent of the 200 million inhabitants of the region. Today, Protestant faiths claim 40 million followers (9 percent) of the region's 450 million people, and 16 million of these are Brazilians. Possibly as many as 10 percent of all Brazilians are now Protestants.

The sudden emergence of Protestantism in Brazil also reflects a dramatic shift away from the so-called mainline denominations (Anglicans, Presbyterians, Methodists, Baptists) to evangelical and pentecostal groups. Fifty years ago, evangelicals and pentecostals accounted for about 10 percent of all Brazilian Protestants. Today, 80 percent of all Brazilian Protestants belong to evangelical and pentecostal faiths. According to one recent study, evangelical Protestants opened five new churches each week in Rio de Janeiro between 1990 and 1992. This explosive shift has been triggered in part by the successful efforts of North American missionaries. Like their North American brethren, the

Brazilian evangelicals and pentecostals stress the direct presence of the Holy Spirit and communication with Him in their church services. They believe in the "speaking of tongues" and miracles through divine intervention. As in the United States, church services are lively, noisy, and emotionally charged.

For some time now, social scientists have noted the strong similarities between the beliefs and activities of these Protestants and Afro-Brazilian and spiritist religious practices. Arguably, the long and profound exposure to Afro-Brazilian religions has made evangelical and pentecostal practices less foreign to Brazilians than they would be to many Latin American Catholics. In both types of religions, the direct presence of God (or gods), spiritual possession, speaking in tongues, and ecstatic religious experiences play a central role. Some pentecostals, despite their antagonism toward the Afro-Brazilian cults, have accepted the validity of the ecstatic experiences of the Afro-Brazilian churches. The root of their bitter antagonism toward these cults is the belief that the cults convene with the devil as well as with God. Practitioners of *macumba, umbanda,* and *candomblé* in Rio de Janeiro, for example, have complained loudly of harassment and persecution by pentecostals who accuse them of worshipping the devil (Exu).

Although Protestantism has made inroads across regions and classes, the *favelas* of major urban areas (especially Rio de Janeiro and São Paulo) have become the principal battleground for the hearts and souls of the faithful. These aggressive Protestant faiths have had their greatest success among poor urban migrants. Many observers have suggested that the new religions offer the urban poor a means of coping with the dislocations and frustrations of moving from the countryside to impoverished urban slums. The evangelicals and pentecostals promise the "followers of the new faith" wealth and success through a profound belief and the traditional virtues of Calvinism: discipline, asceticism, and hard work. The goal is not to destroy the system, but to succeed within it. As with all Protestant sects, salvation depends entirely on faith. Success and "good works," however, are worldly evidence of one's membership in the "elect."

Increasingly, the message of Protestantism appeals to Brazilians, and that message has spread rapidly in recent years with the rise of Brazil's own version of televangelism. In 1950 Brazil had less than 2 million Protestants. That number rose to nearly 9 million in 1980, and from there has nearly doubled in the last decade. More than 500,000 Brazilians join Protestant churches each year. Television and radio evangelism, to a large degree, has helped produce this dramatic increase. A decade ago, Protestant religious programming came largely from the United States with translation or subtitles. In the past few years, a number of prominent Brazilian evangelists have emerged with their own programs, radio stations, and, now, television networks. Twenty years ago, Edir Macedo worked as a civil servant for the state of Rio de Janeiro. In 1977, he founded his own pentecostal church, the Universal Church of the Kingdom of God, housed in an old funeral home. Today his sect has 850 churches and 500,000 members. In April of 1990, he purchased one of the oldest television networks in Rio de Janeiro, and he has begun to rebuild it into a Christian broadcasting system.

Clearly, the dramatic emergence of evangelical and pentecostal Protestantism now presents a profound challenge to the Catholic Church in the world's largest Catholic nation. Those who, in the past, sought to maintain the status quo and remain socially and politically passive must now rise to the challenge of this politically conservative and socially active religious movement. The social and political activists must reevaluate their movement to find out why so many of the poor have chosen the conservative Protestant faiths instead of liberation theology. (As one minister wryly remarked, "The irony is that the Catholics opted for the poor, and the poor opted for the Evangelicals.") A single decade of dramatic growth is certainly a fragile basis for long-range predictions, but Catholics and Afro-Brazilians face a momentous challenge to their long-standing dominance of Brazilian religious culture. It is much too early to tell, but these aggressive Protestants have the potential for fundamentally reshaping Brazilian religious life and culture in the coming years. They clearly believe that they are forging a Brazilian Reformation.

FAMILY, GENDER, AND SEXUALITY

Despite the recent growth of Protestantism, Brazil remains a society profoundly shaped by a Catholic ethos. The ideals of organic social unity, hierarchy, male domination, family, and procreation have been preached and promoted by both church and state for centuries. In this

Iberian Catholic vision, God commanded the universe, kings reigned over their subjects, men ruled over their families, and women served their men as good wives and mothers. As in many traditional societies, the family became the basic social unit, the foundation of a patriarchal social order.

Ideally, the family included not only husband, wife, and children, but also relatives on both the mother's and father's sides, as well as across several generations. While this extended family (or clan, *parentela*) may have been the exception rather than the rule, it was very clearly the ideal of the affluent and the powerful. The ability to construct a clan across generations and through intermarriage with other families has long been a mark of power in Brazilian society. Small or fragmented families are a sure sign of lack of resources (political or economic), and large extended families a visible demonstration of power.

The Iberian practice of long, multiple surnames reflects this exceptional concern about family ties. Your identity depends on who your relatives are and to whom you are connected, and those who cannot demonstrate family lineage surely are unimportant. Like the serfs of the Middle Ages, slaves and the rural poor had no need for surnames, for they had no familial power to display. Those who aspired to be like the elite attempted to emulate their families and their concern for kinship. Only in the last century have all Brazilians been given the dignity of full names. Like other Iberians and Iberoamericans, Brazilians of all classes (especially the middle and upper classes) continue to put great emphasis on names as a way of placing an individual (and his or her importance) through kinship networks. Unlike North Americans who normally retain only the surname of the father, Brazilians typically have at least two surnames to indicate the family of both father and mother. In this male-dominated society the father's surname comes last. Maria Soares Ribeiro (to take a hypothetical example) is descended from the Soares family through her mother, and from the Ribeiro family through her father. Should she marry, she would add her husband's final surname to her own and drop Soares. Her children would have the last names of her father and her husband. Despite the concern for tracing family through both the father's and mother's lineages, the bias is clearly patrilineal.

Ritual kinship reinforces and builds on the biological kinship that binds the family network together. In this Roman Catholic society, godparenting *(compadrio)* has long been an essential social institution.

Parents try to choose allies and protectors as godparents for their children when the children are baptized. Among the elite, the choice of godparents serves to reinforce powerful social and family networks. Those of lower social standing often attempt to persuade the more powerful to serve as godparents. Generally, however, parents turn to those of their own social world to serve as godparents.

Family networks often expand (particularly among the affluent) through the custom of treating associates and dependents as kin. In pre-twentieth-century Brazil, powerful landowners often informally adopted the families of servants and attendants. This custom is still practiced in rural areas. At times, ties are established through ritual kinship, as servants and subordinates turn to the patron *(patrão)* as a godparent. More often, the powerful patriarchs simply assume a paternalistic responsibility for their distant relatives, servants, and associates (known as *agregados,* "the attached ones").

Both biological and ritual kinship continue to bind families together in contemporary Brazil. Comprehending the historic and continuing influence of the extended family in Brazil remains critical to any understanding of Brazilian society. For centuries, the family has been the basic social institution and the foundation of economic activity, social organization, and political networks. Even today, many of the major companies and economic groups in Brazil are family-owned and operated, a pattern that has become rare in the industrial economies of the North Atlantic community. Family networks still exert a powerful influence on the success or failure of prominent social and political groups in Brazil.

In Western society, the strength and integrity of the extended family has been disintegrating for centuries. Maintaining the nuclear family, much less the extended family, has become very difficult in late-twentieth-century Europe and North America. The extreme individualism of North American culture has accelerated the fragmentation of the family, as have the rapid changes spurred by urbanization and industrialization. In the United States, even the most forceful advocates of "the family" generally are referring to the nuclear family. Even among those who consider themselves socially and politically conservative, being pro-family rarely includes residing with or near parents and grandparents. Few families in this country include more than two generations (parents and children) or extended family members living in the same home or in close proximity.

Despite urbanization, industrialization, and modernization in Brazil, the strength and integrity of the family (among the middle and upper classes at least) remain strong. As a rule, Brazilians keep families together by living in close proximity, visiting often, and constantly helping each other out. Generally, single sons and daughters continue to live with their parents even after they become adults. Although the extended male-dominated family developed and evolved in a rural society, it has survived and continues to thrive in modified form in a modern, urban, and industrial Brazil.

Like the extended family, sexual attitudes and behavior have evolved since the colonial era. The American colonies inherited the patriarchal values and attitudes of the European colonizers. In the devoutly Catholic culture of Iberia and Iberoamerica, male domination and female submission were fundamental features of the social and cultural order. Females rarely had access to education, political power, or economic opportunity. As in other European societies, a "respectable" woman could become a wife and a mother, but little else. In addition to her duties as a mother and wife, a woman had to maintain a very low profile in Portuguese society. While in most of Europe she was to be seen and not heard in public, in Portugal she was rarely seen.

Men not only had the dominant role, but they also were not held up to as high a moral standard as women. Premarital sex and adultery (particularly in the higher social classes) could destroy a woman's honor, raise the specter of violent retribution by husbands and brothers, and end all chances for a "normal" social existence. It was expected that men would stray from celibacy and marital fidelity— generally with few negative consequences. Women were expected to emulate the Virgin Mary and maintain a cult of virginity. Men followed their hormones and a cult of virility.

The enormous demographic imbalance between the sexes, along with slavery, intensified this double standard in Brazil. Few white women immigrated to Brazil, and Portuguese domination of Indians and African slaves presented white males with ample opportunity to find sexual satisfaction among Indian and African women. By the nineteenth century, the large racially mixed population of Brazil offered living testimony to the extent of sexual relations between white males and African and Indian females.

The masters did not encourage the formation of stable family relations among the slave majority. Business and profits took precedence over conjugal ties. On the eve of emancipation in the 1880s, less than 10 percent of the adult slave population was married. In the slave quarters and among the free poor, marriage in the church was a rarity. Among the masses, consensual unions became more common than legal marriage, a situation that was true in most of Latin America. Much more so than in North America, sex outside of formal matrimony became commonplace, out of necessity for the masses and out of convenience for elite males.

Among the poor, illegitimate births became commonplace and held little social stigma. Among the elite, the sexual dalliances of males produced illegitimate children, who were incorporated into the family as secondary members. Although not full-fledged family members, these "natural" children (as they are called in Spanish and Portuguese) were recognized and brought into the already extensive family network. The multilineal Iberian family became multileveled.

Colonial Brazil became notorious for its relaxed sexual ambiance. Crown and church officials complained endlessly of the openly permissive behavior of the Portuguese colonists and the clergy. "There is no sin south of the Equator," became an oft-repeated observation. The white male colonists gradually developed a preference for the mulatta over black or white females. Even today, the broad hipped, fleshy mulatta remains the almost mythical symbol of Brazil's racial mixture and the stereotypical female sex symbol. The mulata's fame and desirability reveal a great deal about the underlying nature of sexual, social, and class relations in contemporary Brazil. In Alma Guillermoprieto's recent book on carnival, *Samba,* she cogently summarizes the multi-layered meaning of this sex symbol in a passage that merits extended quotation.

> *Mulatas* are glorified sex fetishes, sanitized representations of
> what whites viewed as the savage African sex urge, but they are
> also, of course, tribute and proof of the white male's power: his
> sexual power, and his economic power, which allowed him to
> wrest the *mulata*'s black mother away from her black partner. At
> the same time the *mulata* serves to perpetuate one of the myths

that Brazilians hold most dear, that there is no racism in Brazil, that miscegenation has been natural and pleasant for both parties, that white people really, sincerely, do like black people. In fact, the aesthetic superiority accorded to light-skinned black women—*mulatas*—underlines the perceived ugliness of blacks before they have been "improved" with white blood. The white skin also serves to lighten a sexual force that in undiluted state is not only threatening but vaguely repulsive, and at the same time, the myth goes, irresistible.

In this century, the Brazilian elite male continues to hold to the traditional double standard, although social conventions do not allow the open displays of disregard for marriage vows and recognition of illegitimate children. The poor, however, continue to rely on cohabitation and are less likely to legalize their consensual unions than those of higher social status. In all social classes, traditional male dominance retains a powerful cultural hold.

As in many Catholic countries, "crimes of passion" by men have long been pardoned legally or de facto, a reflection of the sexual double standard. Men who kill or abuse their wives in the belief that their spouses have been unfaithful have rarely been prosecuted. Although in recent years feminist groups have successfully pressed for the passage of legislation to end this old tradition, crimes of passion against women— and the reluctance of juries and jurists to punish male perpetrators— remain a part of modern Brazilian society. The traditional view that men should dominate and rule, and that women should submit and accept, still pervades Brazilian society.

Known as *machismo* in Spanish, this overweening male pride is characterized by an emphasis on the aggressivity of the male and the passivity of the female. Common to most Western societies, *machismo* is more complex and difficult to pin down in Brazil than in Spanish America. In its classic form, *machismo* stresses diametrically opposed male/female roles. Men are tough, aggressive, and worldly, and destined to rule family, community, and nation. Women are weak, passive, and ignorant, and born to bear and raise children, and care for the family and the home.

In Spanish America, this rigid division in roles leads men to pride themselves on virility as demonstrated by sexual conquests, numerous offspring, and heterosexuality. Men shun any displays of femininity, and the worst possible insult is to question a man's heterosexuality. To call a man a *maricón* (queer) in Spanish is to challenge his very being. The failure to retaliate for such a slur can be devastating for a man's honor. Consequently, homosexuality is not only considered deviant behavior, it is also confronted by severe cultural repression.

Although *machismo* also characterizes sexual relations in Brazil, it takes on a nuanced form that makes it more difficult to pin down than in other Latin and Latin American cultures. While the sexual double standard persists and men pride themselves on virility and sexual prowess, the almost-manic emphasis on heterosexuality and antagonism toward homosexuality and bisexuality are not so pronounced in Brazil. While they still suffer persecution and abuse, gays and lesbians are a visible and accepted part of contemporary Brazilian society. Gays and gay rights movements are a visible and vocal part of the culture of the large metropolitan areas, especially in São Paulo and Rio de Janeiro. Homosexual characters appear regularly in movies and on Brazilian television, more so than in the United States or Spanish America. Certainly homosexuals in Brazil are not perceived as threatening to male sexuality and the social order as they are in the rest of Latin America. As with race relations, the ambiguities and contradictions in sexual relations are striking. Homosexuality is a widespread and notable feature of Brazilian society, yet this has not meant the end of the repression of homosexuals. (According to some gay rights groups, hundreds of gays and lesbians have been murdered during the past decade alone.)

Brazilians' attitudes toward nudity and the human body also differ drastically from those of North Americans and many Latin Americans. In the United States, to pose nude has long had a negative impact on the career of mainstream artists and actors. Witness how often major actresses attempt to suppress nude photographs taken before they became established. In Brazil, many major female actresses have posed nude for men's magazines, a move that is considered simply one part of her efforts to build a career and a following. Xuxa (Maria da Graça Meneghel), who has become Brazil's biggest children's television and film personality in recent years, posed nude on numerous occasions earlier in her career as a model and actress.

Nudity is also more generally accepted than it is in the United States. Anti-pornography lobbies in the United States have pushed magazines with nudity behind the counter in many places. In Brazilian cities, newsstands (on virtually every corner it often seems) are plastered with magazine covers featuring nude males and females. Nudity has also become a prominent feature of television programming. The nightly soap operas often compete to see which can be more explicit. One recent hit series featured a completely nude actress swimming under water as its transition before and after each commercial break. Significantly, the focus is on the female nude, which is another sign of male domination in society.

As a rule, Brazilians are simply less inhibited and uptight about the human body and physical contact than peoples of the developed world. The string bikinis (*tanga*), for example, for which Brazil is famous, are worn by women of all shapes and sizes. The willingness to expose flesh, whether one's physique be fit and trim or fat and flabby, reflects a people much more comfortable with the human body. This "laid back" attitude, however, does not stop the wealthy from resorting to cosmetic surgery. Rio de Janeiro has become the world capital for plastic surgery, and Ivo Pitanguy is its best-known plastic surgeon. The affluent are obviously not completely comfortable with physical imperfections.

Brazilians are also a much more expressive and physical people than North Americans and Northern Europeans. When friends meet, they invariably greet with physical contact. Men shake hands and embrace. Women exchange kisses on the cheeks and hug, as do both men and women when they greet the opposite sex. The warmth of the touching and its duration reflect the degree of friendship and familiarity. Like so many Latin peoples, Brazilians also have a sense of personal space that often seems nonexistent to North Americans. Many times I have seen a Brazilian (literally) back a North American around a room as the latter tries to move away from the former. Brazilians want to be close and touching. North Americans want to keep more distance.

Public physical contact between persons of the same or opposite sex also differs from North American practices. In the past, touching between males and females was limited, reflecting very traditional gender roles. Over the past few decades, as European and North American influence have helped break down some of the traditional male/female role patterns, men and women now touch, embrace, and kiss uninhibitedly in public with no more stigma than would be found in U.S. or European societies. What has persisted, however, and what is absent from North American society, is frequent physical contact between members of the same sex. Much more frequently than in the United States, one sees men touching and embracing. Women often walk arm in arm or hand in hand. In the United States, this type of contact has become increasingly associated with homosexuality. In U.S. society, which already had fairly limited public physical contact among members of the same sex, this has made such contact even less frequent.

Although the sexual preferences and practices of Brazilians have not been as thoroughly studied and surveyed as in the United States, it would appear that Brazilians have a greater tendency toward acknowledged bisexuality. Perhaps this explains the greater acceptance, and perhaps a greater incidence, of homosexuality in Brazil. Males who consider themselves virile and macho often do not see anal sex with other males as "unmanly." As one Brazilian rather bluntly told a foreign writer, "As long as you are doing the fucking, you're not a homosexual. You're *machista*." Bisexuality may serve as a means of blurring the distinction between male and female just as the mulatto blurs the distinction between black and white. And just as with racial mixing, homosexuality may be widespread, yet still repressed.

The bisexual practices of Brazilians have also intensified one of the great tragedies of our times—AIDS. Brazil is rapidly becoming one of the major focal points for the spread of the epidemic, partly because the disease has spread quickly among heterosexuals. It has not been isolated among homosexuals. Heterosexuals currently account for about 20 percent of all AIDS cases in Brazil. Males contaminated with the virus from use of intravenous drugs or gay sex have spread the disease among women. In 1983, just 1 in 32 AIDS victims in Brazil was female. By 1991, the figure was 1 in 6. In 1992, one-third of all newly reported victims were women. The number of deaths from AIDS in Brazil has risen into the thousands in the past few years and is accelerating. With 40,000 registered cases, Brazil now has the third highest incidence of AIDS in the world. Officially, the government says the number of cases registered since 1981 is about 60,000, although some experts calculate the number of those infected with the HIV virus at 500,000. More than 20,000 Brazilians have died from AIDS since 1983. In the state of São Paulo, the number of people infected with the AIDS virus rose from 1.4 per 100,000 in 1987 to 26.2 per 100,000 in 1993, a sure sign of the epidemic's rapid spread. Nearly 50 percent of all cases are in the Rio de Janeiro and São Paulo metropolitan areas.

Two important factors have facilitated the spread of AIDS. The blood supply in Brazil has long operated with little regulation, and it has quickly become a major mechanism for the spread of the virus. In 1987, for example, nearly 20 percent of all cases (as opposed to 1-2 percent in the United States and Europe) were contracted through blood transfusions. A large number of prominent Brazilians have died from contaminated transfusions. Hemophiliacs have been especially hard hit. Herbert de Souza (a prominent sociologist better known as Betinho), and his two brothers (one a famous artist and the other a noted musician), have all contracted AIDS in the past few years (and his brothers have already died from the disease). They were hemophiliacs, and from the upper levels of society. For once, at least, the Brazilian elite have not been able to escape a scourge that plagues the masses.

Further complicating matters is the inability of the health-care system to deal with victims of the HIV virus. The costs and difficulties in developed countries are well known. Although the medical community in major Brazilian cities is very sophisticated, the kind of care necessary for AIDS victims is beyond the reach of the vast majority of Brazilians. The public health-care system simply cannot cope. (Even before the emergence of AIDS, only 1 of 3 Brazilians had access to modern health care, and the country has long suffered from a shortage of hospital beds.) As a result, poor AIDS victims often become social outcasts with nowhere to turn for health care. When they are able to find a hospital that will accept them, they are simply placed in isolated wards with no special treatment. The contaminated blood supply and lack of medical care have compounded the effects of the disease.

Sexual promiscuity, combined with traditional patriarchal politics, have also created a tragic crisis that results in a staggering number of abortions and abortion-related health problems. According to a recent survey (1995) by the World Health Organization, Brazil has one of the highest incidence of abortions in the world—some 1.5 million annually (5 percent of the world total), with some 4 million live births. Although the statistics for Brazil are nearly identical to those in the United States, there is one important difference—abortion is illegal in Brazil.

Much like American women before the *Roe v. Wade* decision in 1973, Brazilian women seek out illegal abortions in enormous numbers. Economic status determines the result. Those who can afford it seek out a safe and discreet abortion. Those who cannot (and they are the vast majority) receive inadequate treatment in back alleys and fly-by-night clinics. The results are devastating. Some 400,000 Brazilian women have to be hospitalized each year due to complications from poorly performed abortions, and about 400 women die from abortion-related complications. The casual observer is tempted to blame the lack of an abortion law on the influence of the Catholic Church. While the Church continues to fight any attempt to legalize abortion, it is clear that this human tragedy should be blamed primarily on a political culture dominated by men who have shown little concern for "women's issues."

The "women's movement" in Brazil has been emerging slowly for decades, but has only recently begun to affect legislation and the political process. Females account for more than half the Brazilian population (51 percent), and nearly 40 percent of all adult females work outside the home (compared with 22 percent in Mexico, 33 percent in Argentina, 66 percent in the United States, and 80 percent in Sweden). Much like Afro-Brazilians, few women as a group occupy positions of power and influence. Only 2 percent of all executives are women (versus 6 percent in the United States), and less than 6 percent of all congressional deputies are women. A slow but steady change has been taking place, however, as the number of female-headed households (now at 20 percent) rises, more and more women enter the workforce, and women increasingly run for public office. Traditional gender roles change slowly, but they are changing as women occupy more and more nontraditional roles in Brazilian society.

CARNIVAL: THE WORLD TURNED UPSIDE DOWN

Brazil's contending religious traditions have shaped a culture that borrows and blends traits from Europe, Africa, and the Americas while also forging new, distinctly Brazilian cultural forms. While Afro-Brazilian cults best illustrate the impact of the interpenetration of religions in Brazil, carnival (carnaval) stands out as the most vivid, striking, and widespread form of cultural blending. Along with coffee and soccer, this ostensibly Catholic pre-Lenten celebration with its bacchanalian revelry has become indelibly linked with the very essence of Brazil. Carnival brings together the starkly contrasting yet deeply intertwined dualities of Brazilian life: rich and poor, African and European, Catholic and Spiritist, male and female.

Many anthropologists have noted that Brazilian carnival is a ritual of reversal that inverts daily reality. Once a year, for four days, carnival offers the illusion that anyone can be anything. This enormous national party releases Brazilians from what some refer to as "the hard reality of life" (a dura realidade da vida) through an illusory, utopian world. Pleasure replaces work; abundance supplants poverty; equality and individuality push aside hierarchy and class. More vividly than any other aspect of Brazilian culture, carnival demonstrates the capacity of Brazilians to experience and express unambiguous joy. More so than North Americans, or even other Latin Americans, Brazilians have an incredible capacity to range from the depths of despair to an almost pure ecstasy. For one brief moment each year, carnival liberates Brazilians from what Roberto DaMatta has called the "web of obligatory social relationships." Connections to family responsibilities, the social hierarchy, and sex roles can be severed in a world turned upside down. Each and every Brazilian has the opportunity to live his or her fantasy (fantasia).

This ritual of reversal allows Brazilians to lower social and sexual barriers and tensions for a brief period before going on with their daily lives. The social side is the easiest to explain. In a society where two-thirds of the population normally lead lives in desperate poverty, carnival provides them with a brief moment of euphoria and fantasizing. The masses, quite literally, take the streets. The scene on the streets of Rio and other major cities is an astonishing, and sometimes frightening, one. Police repression eases and the underclass becomes bolder. The energy and emotion unleashed by the masses on the streets of major urban centers, especially Rio, during carnival is impressive. Many commentators have pointed out how this short, explosive release of social tensions serves as an effective mechanism for maintaining the inequalities and social hierarchy for the rest of the year. Ian Buruma's observation about the Japanese obsession with violent fantasies in film and print also accurately describes one of the social functions of carnival. "Encouraging people to act out their violent impulses in fantasy," he notes, "while suppressing them in real life, is an effective way of preserving order."

Explaining the relationship between sex and carnival presents more difficulties. Increasingly, the sexual side of carnival has become more pronounced and graphic. In the parades, topless female dancers have become commonplace, and full female nudity (although the women do wear body paint) has made its appearance in recent years. In the parties (including those on national television), dancing and revelry are sexually graphic and provocative. It is the female who is on display and not the male, and this reflects the essence of this male-dominated society. These women have become the ultimate sex object in Brazilian society. The fleshy, sensuous, erotic mulata of carnival represents the ultimate sexual fantasy, especially of the elite, white Brazilian male. The evolution of this aspect of carnival demonstrates vividly the subversive power of the Afro-Brazilian past. The sexuality of this idealized woman of African heritage has become central to the celebration of this ostensibly Catholic religious ritual.

Another striking feature of the sexual side of carnival is the inversion of gender roles. For centuries, and originating in Europe, men have traditionally dressed up as women during carnival. Just as the social hierarchy turns upside down, sexual roles reverse. The transformation of men into women through this ritual reflects the widespread (relative) tolerance of homosexuality and bisexuality in Brazil. Being gay or lesbian may be outside the norm, but it is more tolerated than elsewhere in the Hispanic world, and nowhere is this more evident than carnival. For decades, transvestites have played a prominent role in the celebration. *Manchete,* a weekly magazine that is similar to *Life* in the United States, dedicates two entire issues to carnival each year. The dozens of photos graphically display the nudity and sexual side of carnival, and they always include an extensive section of photographs of the transvestite queens of carnival. In a culture with a very relaxed attitude toward sex and sexuality, carnival has become the maximum expression of that attitude. Carnival, observes Roberto DaMatta, "is the glorification of what goes on *below the waist.*"

Despite its close association with contemporary Brazilian culture, carnival has only emerged in its present form during the past few decades. Like Afro-Brazilian religions, carnival has evolved over centuries, borrowing from and blending European and African cultures. Carnival originated as several days of festivities prior to the beginning of Lent, the period of fasting and penitence prior to Easter. A ritual recreation of Christ's wandering in the desert prior to his entry into Jerusalem, and his eventual crucifixion, Lent is meant to be a period of penitence, reflection, and self-discipline.

The origins of carnival date back to medieval festivities of celebrating and enjoying the final days before the onset of Lent. In Portugal, these activities were originally known as the *entrudo*. On both sides of the Atlantic, the Portuguese staged mock street battles in the days before Lent with opposing groups throwing eggs, water, and flour at each other. In early-nineteenth-century Brazil, these battles became so destructive and violent that the government attempted to ban them. Simultaneously, Africans had begun to celebrate by dancing and playing music in the streets. The white elite viewed these festivities with disdain and horror at the openly sexual overtones of the dances.

Gradually, in the late nineteenth century, the music and dancing became widespread and accepted, and more ritualized. The musicians normally played drums, whistles, and rattles, and the *lundu* (a dance of Angolan derivation) evolved into the *samba*. One writer has described dancing the samba as a form of physical schizophrenia. The dancer remains virtually immobile from the waist up, while executing a frenetic and dazzling series of complex dance steps. In the early twentieth century, especially in Rio de Janeiro, dance clubs were formed to give greater coherence and structure to the street celebrations. This process converged with the turn-of-the-century elite move toward elaborate costume balls and dances during the three days preceding Lent. By World War II, the basic structure of modern carnival had taken shape: samba "schools," structured parades, and elaborate costumes *(fantasias)* around a common theme.

In the postwar period, this pre-Lenten celebration has become a national celebration attracting worldwide recognition. The festivities in Rio de Janeiro, in particular, have become an international event attended by the jet-set and tens of thousands of foreign tourists. In the early eighties, the city government (due to the enormous logistical problems of setting up and dismantling the parade route), moved the parades from the central Rio Branco Avenue and built the *sambódromo* on a side street in downtown. The sambadrome looks like the world's longest football stadium (700 meters long) without any walls or seats at the ends, and it holds 65,000 people. Permanent bleachers, box seats, and suites for viewing the parades of the samba schools line the street for several blocks.

The parades have become the visual centerpiece of carnival—and they have become big business. The schools all originated in the slums of Rio and, in conjunction with the Rio Tourist Bureau, have staged the parades on the three nights prior to Ash Wednesday. The schools are year-round social clubs that gear their activities toward the climax of the parade. With membership in the thousands, the big schools design costumes and floats, practice their dancing, and develop a new samba song each year in preparation for the big moment. Until the 1970s, the schools and the parades were largely filled with the poor, non-white lower classes from the slums. More recently, many middle-class whites have joined the clubs and parade with their favorite samba school.

In the 1970s and 1980s, these schools became very big business for two reasons. First, major figures in the numbers game (the *bicheiros*) began to support and gain control of their favorite schools as a means of gaining recognition and prestige. Second, and more important, the parades themselves became a nationally televised event that attracts some 40 million viewers around the country, and the federation of schools sells the television rights for millions of dollars. As is true in professional sports in the United States, television contracts generate enormous revenues that are plowed back into the teams/schools. (The big schools spend as much as $750,000 preparing for the annual parade.) Television has transformed a street festival into a stylized performance for which schools now produce parades tailored for the spectator, both in the grandstands and at home.

Judging these extraordinarily elaborate parades has also become incredibly complex and contentious. On successive evenings, the fourteen biggest and most important schools parade their costumed dancers down the Avenue of the Marquis of Sapucaí before a panel of dozens of judges, who rate them on music, costumes, dancing, and style. Each school is given ninety minutes to samba down the length of the avenue. The competition among the big schools is intense, and their supporters in the stands cheer them on. Winning the competition is akin to winning a professional sports league championship for the year.

The other, more participatory side of carnival is partying. For four nights, the nation becomes one enormous, continuing party. Beginning at about 9 P.M. and continuing until after daybreak, the elites and masses drink, dance, and carouse. For the masses, carnival becomes one giant street party. The rich gather at their private clubs for elaborate costume balls, where they can mingle free from the teeming masses of the Brazilian poor. The television networks broadcast parties from the most famous clubs throughout the night.

During the day, everyone collapses, regroups, and rests before the revelry picks up again the next night. Beginning on Friday evening and continuing through Tuesday night, the party goes on, and business and government comes to a standstill. Monday and Tuesday are holidays, and most workers are not expected to return to work until midday on Wednesday. As carnival approaches in February, one of the most common phrases heard in offices when scheduling business comes up is, "after carnival" *(depois do carnaval)*. Much of the nation's business and government affairs are put on hold until after this ritual release of national steam.

MUSIC, TELEVISION, AND FILM

Carnival highlights the exceptional talents of a people with rich and diverse musical and artistic traditions. The sounds of samba that dominate carnival showcase but one of the many musical styles in Brazilian popular culture. Over the past two centuries, Brazilians have developed a range and depth of musical styles that reflect their multiethnic and multicultural heritage. The music continues to draw upon African rhythms; the rural, mestizo traditions of the interior; and the classical and popular strains of Europe and the United States. Brazilians listen to classical, folk, pop, rock, jazz, country, and rhythm and blues in numbers that rival the music markets in the major industrial nations. Few countries match Brazil for the range and size of its music scene. (The great Argentine musician Astor Piazzolla remarked in 1982 that Latin American music as a whole had produced little of interest or value, with the exception of Brazil, "which has the richest popular music in the world.")

The vitality of Brazilian music testifies to the tenacity and spirit of Afro-Brazilians and their impact on national culture. Beneath the surface of elite/official culture, millions of Africans and their descendants clung to the remnants of their own musical heritage and forged a popular musical tradition that was neither African nor European nor Native American. While the elite listened to the classical and church music of Europe in the colonial period, the slaves recreated the rhythms of Africa in music and dance. Samba is the most famous result of this cultural creativity. In this century, the masses and the elite have also borrowed from European and North American music. Rock and roll (Brazilian style) has become a staple of the young. More interesting is the development of a thriving jazz scene in the major cities. Here is a musical style created by blacks in North America that now has been adopted by blacks and whites in North and South America.

Unlike the United States, Brazil has almost no "black" music scene. While in the United States there are radio stations directed specifically at the black community, and styles of music aimed at black consumers, no equivalent exists in Brazil. (Certain types of music are clearly identified with Africa or blacks, but the effort to establish a separate Afro or black music in recent years has been a small current in a very rich music scene.) The music of both countries mirrors their race relations. Only in a country with polarized race relations could one find a black music market. In a country with complete racial mixture, there is no black target audience, and no "black" music. Beginning in the 1950s, rock and roll brought black musical influences into the mainstream of mass popular culture in the United States. Today, black singers and musicians have achieved mass appeal among white audiences, and the influence of black musical styles has been profound on the "white" music scene. Despite the interpenetration of styles, the music scene remains segmented and polarized. There are still records and radio stations geared to black consumers and to white consumers. Complete racial mixture, and the cultural mixing that goes with it, have made Brazilian popular music exceptionally dynamic and diverse. It has also made it impossible to segment the musical market along racial lines. Like the races in Brazil, music is a continuum, of styles instead of colors.

Brazil has not just been a recipient of influences from music abroad; it has also contributed to the international musical scene. In the late fifties and early sixties, a new strain of samba, sometimes called counterpoint samba, emerged. (It should be noted that samba is both a type of dance and a style of music.) This "new wave" samba, or bossa nova, had enormous impact in Europe and the United States. João Gilberto, Antônio Carlos Jobim, and other Brazilian artists toured and recorded in the United States and Europe. Bossa nova became the rage in the United States, and singers like Frank Sinatra incorporated

Brazilian songs into their repertoire. The English version of *A garota de Ipanema* ("The Girl from Ipanema"), by Gilberto and Jobim (lyrics by Jobim and Vinicius de Morais) became an international hit, and it is one of the five most recorded songs ever written. In the hands of a Sinatra, bossa nova became pop music. (The lyrics were also less sexually suggestive in English than in the Brazilian version.) In the hands of Gilberto and Jobim, bossa nova became jazz. While in New York in 1963, for example, they recorded with saxophonist Stan Getz, producing a vintage jazz album of bossa nova songs.

During the 1970s and 1980s, in spite of (or perhaps because of) the censorship and repression of the military regime, popular music became very diverse and sophisticated. The rich and creative explosion in popular music became known as *Música Popular Brasileira* (better known by its acronym, MPB). Jazz, rock, folk, country, rhythm and blues, foreign and national influences cross fertilized. Milton Nascimento (from Minas Gerais) and Gilberto Gil (from Bahia), both black by Brazilian standards, helped popularize a blend of African rhythms, rock, and jazz music. Superstar "white" artists like Chico Buarque, Caetano Veloso, and Maria Bethânia helped create a uniquely Brazilian popular music with influences from Africa, Europe, and North America, as well as Brazil. (Heitor Villa-Lobos [1887-1959] is Brazil's greatest classical composer and one of a handful of Latin American classical composers whose works are performed by major orchestras in Europe and the United States today.)

The diversity of the music and the enormous population of the country has made Brazil one of the major record markets in the world today. Now the seventh largest market in the world, the Brazilian recording industry is dominated by the big multinational corporations from Europe and the United States. Polygram (Dutch/German), Warner/Electra/Asylum (U.S.), and CBS/SONY (Japanese) have signed and marketed the biggest Brazilian musical stars. Roberto Carlos, Latin America's biggest recording artist, regularly turns out an album of popular/romantic songs that sells millions of copies in Brazil and in the rest of Latin America.

In recent decades, Brazil has exported much more than its musical talent. Since the early 1960s, Brazilian cinema has also gained worldwide attention for its creativity and originality. While bossa nova attracted listeners around the world in the early sixties, Brazil's *Cinema*

Novo ("New Cinema") burst on the scene. In 1962, *O Pagador de Promessas (The Given Word)*, directed by Anselmo Duarte, was named best film at the Cannes Film Festival. Throughout the sixties and seventies, a group of young Brazilian directors produced films noted for their probing look at social issues and their cinematic originality.

Although Brazilian film production began in the 1890s, foreign films, particularly American films, have long dominated the industry. Until the government began to subsidize film production in the 1930s, few Brazilian companies could marshal the resources to make films following the studio model in Hollywood. During the 1940s and 1950s, production of feature-length films rose from a handful each year to more than three dozen, a figure that pales in comparison with Hollywood's 300-400 films per year during the same period. In Latin America today, only Brazil, Mexico, and Argentina have a reasonably strong national film industry. Yet, as is true in the rest of the region, foreign films (75-80 percent American-made) continue to dominate Brazilian theaters. (Brazil is the largest film market in Latin America and the seventh largest in the world.)

In some ways, the emergence of Cinema Novo was a reaction to this foreign domination. Brazilian directors sought to bring to the public their own national reality, to raise the consciousness of the public about Brazilian issues and problems. One of the most prominent of the Cinema Novo directors, Nelson Pereira dos Santos, probed the lives of children living in urban slums in *Rio 40 Graus (Rio 40 Degrees*, 1955). In *Vidas Sêcas (Barren Lives*, 1963) he graphically portrayed the terrible poverty and despair of life in the arid backlands of the Northeast. The work of dos Santos reflects the preoccupation of the Cinema Novo directors with national social problems.

Under military rule (1964-85), the government began to play an even larger role in the film industry through subsidies and censorship. In 1969, the government created Embrafilme, a state company to oversee the film industry. Originally created to promote and market Brazilian films overseas, Embrafilme in the 1970s became the country's major force in film production. The agency became a major source of financing for films, and it shaped the industry through its support for (or antagonism toward) projects. Embrafilme became both a blessing and a curse for the film industry. Although it boosted the industry through financing and support for many projects that would never have been commercially viable, many felt that Embrafilme had become too powerful and restrictive. In the 1980s, the agency became notorious for its mismanagement and bureaucratic rigidity. Many filmmakers criticized Embrafilme for its domineering control of production and distribution. The dismantling of Embrafilme by the civilian government in 1990 has ushered in a new era in filmmaking, with less control from above, but also less financial support from the government. Rising film costs and the end of government subsidies hit the film industry hard. Brazilian feature film production fell from an average of 100 films per year in the late seventies to less than 10 in 1991.

The generals who created Embrafilme clearly saw it as an instrument for creating and controlling cultural production. The power of the agency, in effect, gave it the ability to censor through denial of financing or blocking distribution of films. The Federal Police reviewed all films, foreign and domestic, and rated them. This not only kept any films considered politically unacceptable out of distribution, but it also made the government the guardian of national morality. Films like *The Last Tango in Paris* (Italy-France, 1972) and *In the Realm of the Senses* (Japan, 1976) were banned from distribution in Brazil by the censor. As a consequence, during the 1970s the Brazilian film industry lost the dynamism and creativity of the Cinema Novo years. As many filmmakers backed away from politically dangerous themes and social issues, others began to produce adventure, comedy, and action films. A genre often labeled *pornochanchadas* (comedies with lots of sexual material) became the dominant product on the market by the early seventies.

With the easing of military rule and the transition to democratic government in the 1980s, filmmakers have once again become more probing, more diverse in their subject matter, and more creative. Some of the films of the eighties have been the most successful in Brazilian history. *Dona Flor and Her Two Husbands* (1980), based on an exuberant novel of romance by Jorge Amado, became an international success and the biggest box office hit in Brazilian film history. The female lead, Sônia Braga, has since gone on to success in Hollywood, starring in *Kiss of the Spider Woman* (1986) and Robert Redford's *The Milagro Beanfield War* (1988). Hector Babenco, an Argentine living in Brazil, also burst onto the international scene with *Pixote* (1982), a graphic tale of the life of a small boy in the slums of São Paulo. *Pixote* won the award for best film at the Cannes Film Festival, and Babenco has gone on to direct major

Hollywood productions such as *Kiss of the Spider Woman* (1986), *Ironweed* (1987), and *At Play in the Fields of the Lord* (1992). The 1980s marked the reemergence of Brazilian cinema from years of censorship and control, and its maturation as a creative force internationally.

The most dramatic shift in Brazilian popular culture during the past twenty years has been the emergence of television. In the brief span of two decades, Brazil has become a nation of television viewers and the fifth largest television market in the world. In a country where 20 percent of the population remain illiterate, and daily newspaper circulation barely reaches 8 million copies, the dependence of Brazilians on radio and television as their primary sources of information is even more marked than in the United States. In 1960, fewer than 10 percent of urban homes had a television set, and there were no national broadcasting networks. Today, three-fourths of all Brazilian homes have a television set, and four major networks span the nation. Brazil has more televisions sets (approximately 30 million) than the rest of Latin American combined. Only the United States, Japan, and the United Kingdom boast more sets. Television reaches virtually every Brazilian on a daily basis.

The television revolution has been a direct result of the policies of the military regime in the 1960s and 1970s. The generals recognized the power of television to create a national culture and to serve as an instrument of propaganda. Very quickly after taking power, the military put into place the necessary legal mechanisms to promote broadcasting, and they helped promote the construction of the technological system to make national broadcasting possible: the power system, microwave transmitters, and satellite communications. In an effort to maintain control over this powerful new instrument, they handed out broadcasting licenses to their friends and allies. Support for the regime, or at a minimum, acceptance of the status quo, became a prerequisite for success in the television industry.

The major beneficiaries of these policies were Roberto Marinho and the Rede Globo (Globe Network). An astute, conservative businessman and publisher of one of Rio de Janeiro's major dailies (*O Globo*), Marinho founded Brazil's first television network in April 1965, one year after the coup. Through luck, political influence, and shrewd entrepreneurship, Marinho has constructed a telecommunications giant that has become the fourth largest commercial network in the world.

Globo has more than ninety affiliates throughout the country (30 percent more than his nearest competitor), and it captures the attention of two-thirds to three-quarters of all Brazilian viewers on any given night. Some of its evening soap operas have attained close to 100 percent of the viewing audience during prime time.

Globo has become a major force in Brazilian national culture. The network has been harshly criticized for its support of the military regime and its close ties to the traditional power structure. With the political liberalization of the mid-eighties, Marinho (who personally oversees operations) chose to spotlight the massive popular movement for direct presidential elections. Globo's coverage of the enormous popular demonstrations during 1984 helped make the campaign the largest popular movement in Brazilian history. In the 1989 elections, Marinho threw Globo's immense power behind the candidacy of the young, and then little-known, former governor of the small, northeastern state of Alagoas, and helped make Fernando Collor de Mello the frontrunner and eventual winner of the presidential elections. Globo has the power to influence national opinion and culture far beyond any television network in the United States or Europe.

During the last decade, Globo has begun to face increasing competition, principally from the Sistema Brasileira de Televisão (SBT, Brazilian Television System), created and controlled by Silvio Santos, one of the legendary figures of Brazilian television. Santos' biography reads like a Horatio Alger story. Born Antenor do Senor, this poor child of Jewish immigrants worked as a street vendor until the age of fourteen, when he talked his way into a menial job in the Globo studios in Rio. He has since talked his way to the top of Brazilian television. Within a few years, he had his own program, which eventually evolved into the most popular program on weekend television. Until very recently, the "Silvio Santos Show" (in its various incarnations) ran for hours on Sunday afternoons and evenings. Although the content changed frequently, the pattern remained simple. Santos, the ubiquitous and constantly talking host, moved through game shows, talent shows, variety shows, all in front of a screaming audience dominated by women and youths.

In the mid-seventies, Santos broke with Marinho, and in 1981 he founded SBT. With some forty-eight affiliates, SBT now has about 20 percent of the viewing audience. Although still small compared to the giant Globo, SBT appears to be moving toward a larger and more important role in Brazilian broadcasting. Now in his sixties, Santos has gradually diminished his on-screen presence (doctors have told him he could permanently damage his vocal cords with the continued Sunday marathon shows) and focused on attracting major talents away from Globo. In 1989, he even toyed with running for elected office, and just two weeks before the presidential election he announced his candidacy. A legal technicality prevented him from getting his name on the ballot. Many knowledgeable Brazilian believe that Santos might have been the top vote getter had he been allowed to run. His Sunday program gave him more name recognition (and a better image) than any politician. Such is the power of television in Brazil.

The most popular performer on Brazilian television in recent years has been Maria da Graça Meneghel, better known as Xuxa (shoo-SHAH). She originally became an idol for millions of young girls as one of Brazil's most successful models while still in her teens. Her recent success has been stunning. From 1986 to 1994, she hosted her own children's program on TV Globo for five hours a day, six days a week. The program has become a vehicle for her singing and acting career, and for more than fifty endorsed products. Her records have sold more than 18 million copies (in Portuguese and Spanish). She has starred in eight successful movies, and her comic book (featuring her as the star character) has a daily circulation of 400,000. In 1989, her thirty-nine-city tour sold more than 2 million tickets.

In the last few years, Xuxa has sought a hemispheric audience. She tapes a Spanish-language version of her program in Buenos Aires for export to sixteen South American countries. Univision, the Spanish-language network in the United States, broadcasts the program on weekends for the Hispanic market. During 1993, Xuxa went into syndication in the United States with an English-language version of her show. Filmed in Los Angeles, the half-hour program was carried by more than 200 stations each day. Xuxa's enormous exposure to this hemispheric market has made her the first Latin American to appear on the *Forbes* list of top-earning entertainers. (She was twenty-eighth on the 1992 list, ahead of Harrison Ford and Clint Eastwood.)

In her thirties, blonde, pale, blue-eyed, and of Italian, Polish, and German ancestry, Xuxa is Barbie come to life in the tropics. Her enormous success has generated a lot of commentary about the meaning of a blonde, European idol for a nation that is predominantly non-white. In many ways, Xuxa personifies the complexities of race relations in Brazil. Although seemingly the perfect "Aryan" symbol, she had a highly publicized romance with Pelé in the early 1980s, and this was hardly a pairing that white supremacists would care to see. It is true that the idolization of this blonde, blue-eyed white woman by millions of black and brown Brazilians may convey deep-seated messages about race and power. Nevertheless, in a society as racially mixed as Brazil, and in which other idols of different shades compete for the loyalty of fans, the messages are more complex and complicated than they would be in a racially polarized society (like the United States).

The most popular program in Brazil is Globo's *Jornal Nacional* (JN, National Journal), watched by 60 percent of the Brazilian viewing audience (35 million people) every night at about 8:30 P.M. (In comparison, the leading evening news program in the United States in 1995, ABC's *World News Tonight*, generally had approximately a 15 percent share of the viewing audience.) Cid Moreira, who has been the principal anchor since the JN began in December 1968, may be the best known public figure in Brazil. During the years of military rule, the JN became the nation's major source of news, and under strict censorship it rarely showed Brazilians the most important events taking place: guerrilla activities, torture, repression, and political dissent. With the removal of censorship in 1979, the JN moved timidly to more investigative and critical reporting. This timidity, and the powerful role of the JN, has made it a frequent target of criticism. It is the most influential program on the most powerful network in Brazil. Presidents, senators, and governors, among others, have been known to call Roberto Marinho during the broadcast of the *Jornal* to protest the slant of the coverage.

With Globo leading the way, Brazil has become a major producer of programming, and it exports its programs to more than one hundred countries. In the 1960s and early 1970s, the majority of programming came from overseas (especially the United States) and was dubbed. Today, virtually all prime-time programming is produced in Brazil and has achieved a technical standard equal to that in the United States and Europe. *Telenovelas* (soap operas) have become the quintessential Brazilian program. *Novelas* dominate primetime (6-11 P.M.) Monday through Saturday. Unlike U.S. soap operas, Brazilian *novelas* do not run forever. Most last a couple of hundred episodes, until all the various subplots come to a dramatic conclusion. Competition for the major actors and scripts is fierce, and the productions have become increasingly expensive and elaborate. Imagine taking a major mini-series from U.S. television and extending it six nights a week for a year or so and you have a feel for the nature of the *novela*.

Brazilian networks have long exported their programming to Portugal. In recent years, Brazilian *novelas* (with dubbing) have been big hits in England and on the European continent. During the 1980s, Marinho and Globo bought a television station in Monaco to establish a direct link into the European market, particularly Italy. More so than Brazilian cinema, Brazilian television has the potential to move beyond the national market and to achieve global influence in the coming decades.

LITERATURE AND IDENTITY

Like cinema and television production, Brazilian literature has begun to reach an international audience in the past thirty years. The so-called boom in Latin American literature has moved the region to the cutting edge of world literature. The boom has been the final step in reversing a centuries-old pattern of Latin American writers' striving to emulate and imitate the latest literary trends in Europe. In the nineteenth century, Latin American writers simply adapted European literary genres to local themes. By the turn of the century, they had begun to turn increasingly inward in efforts to define national identity. In the aftermath of World War I, Latin American literati declared their cultural independence and focused on creating their own literary models. The boom marks the coming of age of Latin American literature, as North Americans and Europeans now find their literature influenced by Latin American writers. Although they have not been as universally recognized as their Spanish American counterparts, Brazilian writers have been widely translated and read in Europe and the United States. The one-way cultural flow of the past (from Europe westward across the Atlantic) has now become a mutually beneficial exchange among writers around the globe as Latin American writers have carved out their own niche in world literature.

The greatest writer in Brazilian literature was the first to break away from the cultural imitationism of the late nineteenth century. Joaquim

Maria Machado de Assis (1839-1908) was one of those remarkable individuals who defies categorization or simple explanation. The illegitimate son of a Portuguese maid and a mulatto father, Machado de Assis was largely self-taught. As a teenager, he became a typesetter and proofreader and began to publish poetry, and, eventually, brilliant short stories and novels that made him the most famous literary figure of his day. He founded and served as the first president of the Brazilian Academy of Letters.

The economy of style, wit, and subtle psychological insights in his novels and short stories have made his writings classics. All of his novels have been translated into English (*Epitaph of a Small Winner* [1880] and *Esau and Jacob* [1904], to name but two examples), and he is recognized by literary critics in the English-speaking world to be one of the great overlooked writers of the last one hundred years. Susan Sontag has called him "the greatest author ever produced in Latin America"—no small claim in the land of Jorge Luis Borges and Gabriel García Márquez. Like many great writers, Machado does not fit into any neat categorization. While Machado dealt with universal themes in his work at the turn of the century, most Brazilian writers grappled with ways to apply European models to Brazilian themes, or with ways to break from European influence.

The work of Euclídes da Cunha (1866-1909), offers a dramatic contrast to the urban, cosmopolitan, and universal literature of Machado. Trained as a military engineer, da Cunha worked as a journalist and witnessed the crushing of one of the great social uprisings in late-nineteenth-century Brazil at Canudos in the backlands of Bahia (as discussed in chapter 1). Led by a religious mystic, Antônio the Counselor, the thousands of inhabitants of Canudos fought with local authorities in a series of skirmishes that eventually escalated into a major uprising. The determination and tenacity of these backlanders who wanted to be left alone proved too much for local and state authorities, and they were a tough test of the Brazilian army. A six-month siege and thousands of wounded and dead was the price to crush the rebellion and its leader.

Da Cunha wrote a long epic account of the uprising that combines journalism with social commentary and ruminations on the nature of Brazilian identity. Published in 1902, *Os sertões* (*Rebellion in the Backlands* in the English translation) is a window into the psyche of the

Brazilian intelligentsia at the turn of the century as they grappled with how to reconcile their European fixation with their Native American and African heritage. Da Cunha has enormous admiration for the racially mixed people of the interior, and he recognizes that *they* are the true Brazilians. Yet he desperately wants Brazil to be European, and that would mean the gradual elimination of the racially mixed people of the backlands and their replacement with European immigrants. "We are condemned to civilization," he declared. "Either we shall progress or we shall perish. So much is certain, and our choice is clear." Da Cunha and other intellectuals were trapped in what seemed to be an inescapable dilemma: to be "progressive" and "modern" meant to turn their backs on their own heritage and to stop being Brazilian.

Not until the 1920s did Brazilian intellectuals escape this dilemma by rejecting it. During the centennial of Brazilian independence in 1922, a group of avant-garde artists and intellectuals declared their cultural independence from Europe. These intellectuals argued that they had no desire to be European, they simply wanted to be Brazilian. "Let us forget the marble of the Acropolis and the towers of the Gothic cathedrals," Ronald de Carvalho exclaimed. "We are the sons of the hills and the forests. Stop thinking of Europe. Think of America." In essence, they recognized that "European" was not synonymous with "progress" or "modernity," something da Cunha could not see. They organized a "Modern Art Week" of exhibitions and lectures in São Paulo in February 1922 to announce their break with the past. Led by the versatile *paulista* Mário de Andrade (1893-1945), the Brazilian "modernists" set out to create an authentically Brazilian art, literature, and culture.

For the next thirty years, Brazilian artists and writers experimented and created new forms and models for expressing national culture, drawing on the distinctive features of Brazilian civilization. A group of gifted writers in the Northeast produced a series of innovative novels that brought to life the society of the old sugar plantation region. Graciliano Ramos (1892-1953), for example, in *Barren Lives* (1938), depicted the poverty and desperation of life in the drought-stricken backlands of the Northeast in a stark and spare style.

The most creative and innovative of the so-called regionalist writers, however, did not come from the Northeast, but from the interior of Minas Gerais. Born and raised in Cordisburgo in the hot savanna of the *mineiro* North, João Guimarães Rosa (1908-1967), was trained as a physician. An enigmatic and inaccessible figure, Guimarães Rosa is the James Joyce of Brazilian literature. Much like Joyce, he reshapes the language of the novel, creating, inventing, and reconstructing it. His works are a linguistic tour-de-force, and they are not easy to read, even for the native speaker of Portuguese. His masterpiece, *Grande Sertão: Veredas* (originally published in 1956, and translated into English as *The Devil to Pay in the Backlands*), probes the psyche and superstitions of the outlaws of the interior.

Brazil's greatest poet, and a contemporary of Guimarães Rosa, was also from Minas Gerais. Carlos Drummond de Andrade (1902-1987) came from Itabira, a small town in the old mining zone of central Minas. As a young man, he went to school in Belo Horizonte, and he spent most of his adult life in Rio de Janeiro. His beautiful, simple, lyrical poetry is deeply rooted in the Minas of his youth, yet transforms these regional settings into universal statements on the human condition. (In one of his more famous poems on a legendary episode in the history of Minas Gerais and the nation, he closes with the line, "All history is remorse.") Unlike the major novels of the past thirty years, the poetry of Drummond is virtually (and undeservedly) unknown outside of the Portuguese-speaking world.

The most famous living Brazilian writer is, without a doubt, Jorge Amado (b. 1912), whose books have been translated into many languages and made into films. (*Dona Flor and Her Two Husbands*, the biggest Brazilian box office hit ever in the overseas market, is based on one of his novels.) A native of Ilhéus in southern Bahia, Amado published his first novel in 1932 at the age of nineteen. Nearly all his novels (and he has written a score of them) are set in the Northeast, usually in the city of Salvador, where he lives. Until the 1950s, a strong emphasis on social issues and the harsh realities of life in the Northeast characterized his writing. During the past thirty years, works like *Gabriela: Clove and Cinnamon* (1958), *Dona Flor* (1966), and *Tent of Miracles* (1969) have celebrated the exuberance and sensuality of life in Bahia, the most African and racially mixed of Brazil's large cities (only 20 percent of the population is white). Jorge Amado's work is a window into the mulatto soul of Brazil.

Brazilian literature has come of age. No longer simply imitators of European trends, Brazilian writers produce poetry, fiction, and drama with a complexity and diversity that equals or surpasses that of the developed world. Brazilians continue to have a large and seemingly insatiable appetite for the latest works by European and North American writers, but Europeans and North Americans also now look to the latest works of Brazilian writers.

Yet despite the maturity and quality of contemporary Brazilian literature, relative to publishing markets in the developed world, the Brazilian book market is small. Only in recent years have a significant number of writers been able to support themselves with royalties from their works, and many fine writers still must find work in television, journalism, and other fields to make a living. (The biggest-selling writer in Brazil today is Paulo Coelho, whose books—*The Alchemist* and *Diary of a Magician,* to name but two examples—portray the mystical and magical peregrinations of a man in pursuit of knowledge and wisdom. Written in a vein similar to Carlos Castañeda's *Teachings of Don Juan,* they are seen as fiction by some and non-fiction by his most ardent fans. Four of his works were recently reissued as mass market paperbacks in editions of 1 *million* copies each.) Here the realities of the social question come back to the fore. In a society in which two-thirds of the population live in poverty, the book business becomes a luxury of the elite. (The book market in Brazil is roughly the size of the market in the Netherlands—a country with a population of 14 million—in total sales. The U.S. market—by far the largest in the world—is twenty times larger than that of Brazil.) Illiteracy remains high, and television and radio have become the major sources of information for the masses. In recent years, educators and intellectuals in the United States have lamented the decline of reliance on the printed word, and they have often blamed the visual media. Unlike in the United States, in Brazil the visual media reached a mass audience *before* the achievement of widespread literacy. Rather than having to fight to recapture a lost audience, Brazilian writers must struggle to *introduce* their audience to the printed word. The struggle promises to be a difficult one.

SOCCER: BRAZIL'S SECOND GREAT PASSION

One surefire way to get Brazilians to read is to write about soccer. Football, as it is known everywhere in the world except the United States, is the world's most popular sport, and nowhere else is it more popular than in Brazil. The international soccer championship (World Cup), held every four years, draws some 20 *billion* viewers worldwide compared with about 2 billion for the Olympics. In much of Europe and Latin America, all activity comes to a virtual standstill during the World Cup telecasts. Since the tournament began in 1930, European and Latin American teams have won every championship.

Along with coffee, carnival, and the Amazon, Brazil is best known for its excellence in soccer. Developed in England in the mid-nineteenth century, and spread to the rest of the world though the enormous power and influence of the British Empire, football (or *futebol* in Portuguese) probably arrived in Brazil in the 1880s. The first professional teams were formed at the turn of the century, and professional leagues came of age in the 1930s and 1940s. In 1950, Brazil hosted the World Cup and suffered a stunning defeat in the final game which was played in the newly constructed Maracanã stadium. (With a capacity of 220,000, it remains the largest stadium in the world.)

The excellence of Brazilian soccer over the last fifty years is undeniable. Brazil is the only country to have qualified for every World Cup tournament (fifteen), and it was the first country to win the coveted Jules Rimet Cup three times (in 1958, 1962, and 1970). In 1994, Brazil became the first four-time winner. (Italy and Germany have each won the Cup three times.) Brazil exports many of its best players to the high-paying European leagues, and it has produced the greatest player in the history of the game—Pelé. The Brazilians are famous for a rhythmic, balletic style of soccer that emphasizes finesse and grace. (The Italian filmmaker Pier Paolo Pasolini once said of soccer, "There are two types of football, prose and poetry. European teams are prose, tough, premeditated, systematic, collective. Latin American ones are poetry, ductile, spontaneous, individual, erotic.")

Soccer is virtually the only major league team sport in Brazil. Although basketball and volleyball have professional leagues and have produced world-class players and teams (Brazil regularly wins international championships in volleyball), the teams are few in number and are concentrated in a few cities. Only Formula I racing comes close to generating as widespread excitement as soccer, and it is purely a spectator sport, broadcast almost entirely from overseas. (The great success of Nelson Piquet and Ayrton Senna, who between them have won half a dozen world championships, should certainly prompt some speculation as to why Brazil is the only country outside the northern industrial world to produce Formula I champions. Perhaps it has something to do with the way Brazilians drive—as if they were all Formula I drivers.) There used to be an old saying that there were only two sports in Texas—football, and spring football. In Brazil there is but one sport—*futebol.* A sport that is tailor-made for the masses, soccer requires only a ball (or some facsimile) and some space to play.

In some ways, the sport is also much more democratic than the high-powered, corporate structure of professional sports in the United States. The teams are owned by non-profit clubs that virtually anyone can join, and decisions are made through the representative bodies of the clubs and their directors. To maximize fan identification, the clubs encourage large memberships and low ticket prices. Corinthians of São Paulo has more than 135,000 members. Flamengo (Rio de Janeiro), the most popular team in the country, has some 65,000 members. Ticket prices in the grandstands are kept at around fifty cents and this produces some huge and vocal crowds.

There are more than 8,000 professional soccer players in Brazil, nearly three times the number of athletes in all professional sports leagues in the United States. Rio de Janeiro alone has a dozen professional teams, four of world-class quality (Flamengo, Vasco da Gama, Botafogo, and Fluminense). The soccer season runs all year, with only a three-week interval between the end of the national championship and the beginning of the new season.

As if the overwhelming dominance of the sport and the endless season were not enough, the Sports Lottery (*Loteria Esportiva*) provides Brazilians with yet another opportunity to follow soccer year round. Each week, an estimated 90 percent of all adults (male and female) participate in the lottery by marking their predictions (win, lose, draw) for fourteen games. Two-thirds of the revenue from the lottery goes to pay for social welfare programs run by the government. The other one-third pays the winners, and it often produces payoffs in the millions. The chance to hit "the big one," as with the state-sponsored lotteries in the United States, attracts the bets of the masses each week.

Soccer has also fittingly produced the most famous Brazilian of all times, Edson Arantes do Nascimento, better known by his nickname, Pelé. A black man (by Brazilian standards) from a small town in southern Minas Gerais who did not get past the fourth grade, Pelé is a multimillionaire, and one of the most recognizable figures anywhere in the world today. A member of the first three of Brazil's World Cup championship teams, Pelé is the only player in the history of the game to score more than one thousand goals. His offensive skills are legendary in the sport. Long the star of the Santos club (near the city of São Paulo), he played briefly, after his "retirement" in 1974, for the New York Cosmos in a failed effort to bring professional soccer to the United States. Today Pelé splits his time between Europe, the United States, and Brazil as an international businessman, soccer commentator, and film star. In a survey in the early 1980s, 99 percent of all Brazilians polled recognized his photograph. (Only 88 percent correctly identified the president's photograph.)

Soccer is not just a sport in Brazil, it is a national obsession. It is a common bond (especially among males) that brings Brazilians together, regardless of class or region, especially during the World Cup competition. Just as the United States's fascination with football reflects the violence, intensity, and impersonality of North American society, soccer serves as a metaphor on Brazilian society. Finesse, rhythm, and exuberance are words that describe the style of both Brazilian soccer and Brazilian life. At the same time, much like Brazilian society, soccer is a game in which the masses must be content to watch a few powerful players and teams dominate the play.

To understand soccer is to peer into the soul of Brazil, and into the heart of this lusotropical civilization. This sport that Brazilians have come to love so passionately and play so well reflects the enormous diversity and antagonisms within Brazilian society. A European creation, played by the descendants of Africans and Europeans in the Americas, soccer has created teams, clubs, and fans that mirror the racial mixture and social hierarchy. Fans pray for the teams and make offerings to the ancient gods of Africa to guarantee their success. By the millions, Brazilians bet their savings on soccer in pursuit of quick wealth. The games become brief celebrations with music, dancing, and team colors that momentarily brighten the lives of the masses who live in desperate poverty. Soccer has become a microcosm of this rich, complex society, borne out of the collision of three civilizations and shaped by their conflicts and convergence.

FOUR

Power and Patronage

> Brazil is a nation of pragmatists. Unembarrassed
> by rigid commitment to ideas or principles, the
> Brazilian is a skillful temporizer—tractable
> where others are unyielding, attentive to practi-
> cal consequences rather than to the dictates of
> theory.
> —Frank Bonilla

STRIKING PARADOXES define Brazilian politics. Despite a political spectrum that runs the gamut from extreme right to extreme left, Brazilians are famous for their pragmatism and lack of commitment to ideological principles. The third-largest democracy in the world, Brazil is characterized by a politics that has long been dominated by contending groups of elites. In Brazil, after nearly half a century's experience with modern political parties, personalism and the politics of personalities remain central to national politics. In a political culture that proclaims the modern principles of both individual rights and equality before the law, the state remains an enormous patronage system that rewards the privileged few, reinforces hierarchy, continues to centralize the power of the state over the individual, and often shows little regard for due process. Despite decades of experience with the instruments of democratic politics, authoritarian regimes and military intervention into politics have left deep imprints on Brazilian political culture.

Pragmatism, elitism, personalism, clientelism, and centralism do not make Brazilian politics unique. We can find all these characteristics to some degree in other democratic societies. What is striking about Brazilian politics is that these are guiding principles in a system that theoretically shuns them. Politics in most societies suffer from the gap between stated ideals and the realities of everyday political life. The obvious and clear contrast between "what is" and "what ought to be" makes Brazilian politics fascinating—and frustrating.

TWO POLITICAL CULTURES

The contrasts between the political cultures of Brazil and the United States in many ways reflect the principal paths taken by the nations of the West since the Renaissance: democracy and decentralized power versus authoritarianism and centralism. Although these are not linear and mutually distinct paths, the United States has moved more clearly in the direction of dispersed power while Brazilian elites have been very reluctant to relinquish the power they have concentrated in their hands. In a broad sense, both nations arose out of what, at first glance, appear to be very similar histories. Created by European imperial powers, both nations gained their independence when colonists challenged their European rulers and won. A stated commitment to evolving representative politics has characterized both societies for the past two centuries. Today, the United States and Brazil are the second- and third-largest democracies in the world (after India).

These broad historical parallels, however, mask very profound differences in the development of political traditions in the United States and Brazil. The political heritage of the United States emerged as a branch of the well-rooted traditions of limited government and individual rights that had been growing in England long before the English reached the shores of North America. Brazilians trace the roots of their political traditions back to the statist and centralist political culture of an Iberian monarchy that faced few checks on its power before overseas expansion, or during centuries of colonial rule. Americans are a people with an almost fanatical commitment to the sanctity of individual liberties and a deep-seated suspicion of the power of the state. Brazilians stress the importance of the collective interests of society and have long looked to the state as the ultimate guarantor of security and well-being.

The past does not dictate the present, but it has shaped the evolution of political culture in both nations. Brazilian politics have been molded over the centuries by a medieval, Iberian Catholic tradition that diverged from the path taken by Protestant Western Europe in the aftermath of the Middle Ages. Western Europe witnessed the emergence of checks on the power of monarchy, a growing emphasis on the rights of the individual, and a mechanistic vision of politics. The Reformation in the sixteenth century and the Scientific Revolution in the sixteenth and seventeenth centuries profoundly influenced and intermingled with the rise of the Western European political heritage. These two extraordinary processes, one religious and the other intellectual, contributed to the erosion of the power of established authority in Europe, both sacred and secular.

Powerful monarchs in Spain and Portugal kept the Reformation north of the Pyrenees, and Iberia remained on the periphery of the Scientific Revolution. The medieval Catholic vision of Thomas Aquinas (1224-74)—a vision of hierarchy, order, and community—maintained its grip on the Iberian world. Rather than seeing it as a mass of individuals who somehow managed to come together into a fine-tuned machine, the Iberians envisioned society as a body *(corpore)* in which everyone had a place and function. Individuals knew their place and their function, and by performing their ascribed duties they kept the body politic healthy.

A growing and complex bureaucracy formed the skeleton of this body politic, a bureaucracy in which power flowed upward to the king, who headed the system. Powerful landowners, merchants, military officers, clergy, and public functionaries kept the masses in check by dominating the state apparatus and the patronage it dispensed. The state controlled and dispensed resources (patrimony) to allies and friends and denied patronage to enemies. The state bureaucracy took on a life of its own, certifying, verifying, and acting as the buffer between the powerful and the powerless. This patrimonial system, with its corporatist ethos of hierarchy, stability, and concentration of power, stood in stark contrast to the shifts taking place in western Europe, especially in England, by the sixteenth and seventeenth centuries.

Richard Morse, one of the most astute thinkers on things Brazilian, has long pointed out the powerful imprint that the Middle Ages left on Luso-Brazilian culture. As he has noted, the patrimonial system was pluralistic, but not representative. It compartmentalized privilege rather than promoting equality before the law. The components of society interacted with each other through the paternalistic state rather than relating directly to each other. The result is that Iberians and Iberoamericans have a stronger attachment to the "natural order" and the "human community" than do Anglo Americans. As Morse observes, the belief that "man makes and is responsible for his world is less deep or prevalent" than in other parts of the West. Whereas North Americans have historically delegated power to their leaders, Latin Americans up to the present still prefer to relinquish power to their leaders.

The Enlightenment of the eighteenth century and the Age of Revolution (1770-1830) brought fundamental changes to the political cultures of the Americas, both North and South. In North America, the Enlightenment reinforced the move toward limited government, challenged established authority, and provided the intellectual ammunition for the American Revolution and the forging of the United States. In Latin America, the new ideas also helped produce a crisis of authority, and in Spanish America they stoked the fires of revolt, fueling the wars for independence after 1808.

Rather than reinforcing a long tradition of reform, the new ideas profoundly shattered a deeply entrenched corporatist order, making the construction of the new Spanish American nations especially difficult in the first half of the nineteenth century. Throughout the nineteenth century, the nations of Spanish America struggled to make the shift from a politics built on inherited authority to a contractual authority derived from the consent of the governed. Establishing a new form of political legitimacy proved very costly as factionalism and civil war ravaged most of Spanish America until the mid-nineteenth century.

In Brazil, the influence of the Enlightenment did not produce a dramatic rupture with the corporatist, bureaucratic tradition. The transfer of the monarchy to Rio de Janeiro in 1808 and the unusual role of the Braganzas (both João and his son, Pedro) in providing a relatively bloodless transition to independence helped Brazil avoid the chaos and anarchy so characteristic of Spanish America. In many ways, independence signaled the transfer of power to a Brazilian elite, who continued the process of extending the patrimonial state's power over the vast regions of the empire. In the nineteenth century, the Brazilian elite faced the task of forging a nation out of a colonial beginning, rather than creating a nation by breaking with a colonial past. Continuity rather than radical change characterized independence in 1822.

The task of the successful revolutionaries in both the United States and Brazil was the same—imposing a national government on a sparsely populated Atlantic coastline, and extending its power westward into a vast interior inhabited by hostile Indians and challenging environments. In the United States, a small land-owning (and often slave-holding) elite strove to forge a nation by building on the long-developing British traditions of liberty and limited government. In Brazil, a small land-owning (and largely slave-holding) elite moved to impose centralized government and patrimonial politics, despite a rhetoric of classic liberalism.

In the century after independence, the United States moved gradually and fitfully toward more representative and democratic politics. The move toward a more open form of politics was never easy. More than 600,000 people died in the bloodiest war in the nation's history, a war that freed 4 million slaves and finally gave them the rights of citizenship. Yet, despite the advances, women still had no political voice, and the freed slaves quickly saw their political voice stifled by the imposition of Jim Crow laws in the South. The subjugated Native Americans were isolated and excluded from the system completely, treated as foreigners in their own homeland. By the end of the nineteenth century, the vote and a political voice had been extended, but largely to the white, male masses in the cities and countryside.

For most of the nineteenth century, Brazil's experiment with monarchy made it unique among the nations of the hemisphere. The sixty-seven-year reign of Pedro I and his Brazilian son, Pedro II, provided the country with a stability matched by very few nations in Latin America or by the United States. While civil wars tore apart most of Spanish America and the United States, Brazil's imperial system helped provide a gradual transition from colony to republic. The advantages of stability and gradual change, however, came at a very high price for the masses. Slavery remained intact until 1888, and the free masses had little voice in politics. The poor, rural, non-white masses occasionally expressed themselves through regional revolts that challenged the central government (especially in the 1830s and 1840s). By and large, however, the elites remained firmly in control, and the major crises in the imperial system generally came from intra-elite rivalries.

THE PATRIMONIAL STATE

These elites hold the key to understanding the workings of Brazilian politics over the last two centuries. The ability of various elite groups to keep the masses at bay and maintain control of the reins of power—the patrimonial state and its bureaucratic machinery—has been impressive. Whether under an imperial monarchy, republic, or dictatorship, or through mass politics, elite manipulation of the state machinery has continued, changing outward appearances like a chameleon. In a sense, the state machinery has been the enduring constant in Brazilian political history, while the elites who have controlled the instruments of power have changed throughout the centuries. Sugar planters and merchants gave way to coffee planters, the military, and industrialists, who ultimately had to reach an accommodation with the urban middle classes.

In contrast to Spanish America, in Brazil the Enlightenment and the struggle for independence did not shatter the corporatist, patrimonial system of the Iberian monarchy. The smooth transition to independence with a Portuguese crown prince becoming a Brazilian emperor merely Brazilianized the colonial political culture. Under Pedro II, who saw himself as the Brazilian equivalent of Britain's Queen Victoria, constitutional monarchy replaced the absolutism of the colonial order. The Brazilian elites now had a greater voice in the system, yet it was still a patrimonial political culture. The Constitution of 1824 and subsequent legislation provided for an electoral system and a parliamentary-style politics, yet with a very restricted electorate.

In Brazil, as in most of Latin America in the nineteenth century, classical political and economic liberalism ostensibly triumphed. Yet the apparent triumph of liberalism masked a deep-seated tension that continues to run through Latin American politics. Political leaders spoke the language of liberalism, yet continued to concentrate power in the hands of the few. As Roberto DaMatta has pointed out, a profound tension also arose from the efforts to impose and "institutionalize an individualistic, egalitarian political system in an ideological landscape that is impressively marked by hierarchy and holism." The discourse of liberalism masked a culture of patrimonialism.

Furthermore, until the rise of a small socialist and communist movement in the early twentieth century, no substantial ideological disputes divided the Brazilian elites. Virtually all of them accepted the rules of the system and fought to control it, rather than to change it. The relatively small political elite came from a very homogenous background. Generally from landholding (and often slave-holding) families, they attended the same schools, graduated from one of Brazil's two law schools (Recife and São Paulo), and were generally inculcated with the same constellation of values. Most worked their way up through the political system and followed similar career patterns. Seemingly following a democratic set of rules, the elites in reality were engaged in a form of "shadow play" that hid the true mechanisms of power in Brazilian politics and society.

Politics in modern Brazil has long been a sophisticated game of patronage. In the nineteenth century, one rose through the system by becoming a client of the powerful and by building a clientele among the less powerful. The most important forms of patronage were the spoils of the state. Positions in government became a means to reward one's clients and to advance one's own interests. Brazilian politics became a giant spoils system, and politicians vied for control of the spoils. The state served as the apparatus for accumulating, controlling, and dispensing patronage.

With severely restrictive requirements for voting, and careful scrutiny of the voting booth by local landowners, the electoral system became a means for legitimating the power of the elites. The only real question was which group within the elites would control more of the spoils. Rather than building political parties along the lines of ideology and issues, parties became networks of powerful clans whose political foundations were regionally based. Local bosses became clients of regional political leaders who then contended for power in state governments. The most powerful states (principally Minas Gerais, São Paulo, Rio de Janeiro, Bahia, and Rio Grande do Sul) then jockeyed for position on the national scene. Clans and state elites became the most important keys to understanding the networks of power in Brazilian politics.

Over the past two centuries, Brazilian politics has oscillated between periods of power centralization and decentralization. The strong centralizing tendencies of the imperial state gave way in the 1890s to a more decentralized system in which state governors exercised enormous power and resources. In the 1930s, Getúlio Vargas recentralized power once again; this trend, however, diminished under the populist politics that followed World War II. The military regime reasserted centralism with a vengeance after 1964. With the return of civilian rule in the 1980s, the new regime once again gave enormous power and resources to the state governments. Today, governors exercise powerful leverage in the national political process. Despite the trend toward decentralization in recent years, the Brazilian political system remains highly centralized when compared with the United States. Old habits die hard.

Power continues to emanate primarily from Brasília and the state capitals. The federal and state governments collect taxes and disburse them downward and outward through the bureaucracies to those who support the people in power. The system is incredibly hierarchical and top-down in comparison with the United States. Teachers, for example, are tied into a statewide secretariat of education that controls resources and sets standards. State secretariats, likewise, are tied into the national ministry of education which controls federal education resources and sets national standards. The same could be said of most sectors, from clerical workers to health care.

Brazilian elites have been extraordinarily creative at concocting a curious blend of "representative" government and traditional patrimonial politics to adapt to social and economic transformations during the past two hundred years. At the heart of this creativity has been an ability to absorb new elites into the system and to keep new challenges from below from forcing fundamental social change. What is striking is that throughout most of the past two hundred years, the elites have managed to maintain their power through the manipulation of the electoral process. Brazil has nearly two centuries of experience with some form of elections and representative government. For most of that time, however, elites have controlled or manipulated the electoral process to avoid truly representative, democratic politics. Authoritarianism, whether violent or paternalistic, has long been a key feature of Brazilian political culture.

At the turn of the century, for example, the elites developed a system known as *coronelismo* (literally, colonelism) that beautifully illustrates the tradeoffs and creativity involved in maintaining elite power. At times, the federal government did not have the resources to extend the central state's power into the vast interior. The regional elites wanted to minimize central government intervention into state politics. Large landowners (often with the honorary title of colonel in the local militia, much like in the Old South of the United States) remained very powerful at the local level. The genius of *coronelismo* was in linking up the power of these rural landowners with the resources of the emerging state.

The powerful landowners contended for supremacy in their locale with the state government's support generally proving decisive in any power struggle. The state government could back the landowner with laws and legislation, resources, and, ultimately, with troops. In return, the colonel could guarantee his patrons in the state government the extension of the state's power into the countryside at a minimal cost. He also delivered the vote on election day. The ever-increasing patronage power of the state and federal governments became even more important and reinforced the elites' hold on power. The real struggles in the First Republic were between elite groups for power at the state level. When these struggles threatened the stability of the republic itself, the federal government would send in troops to restore order.

This tradeoff, of state support for large landowners in exchange for the vote, provided Brazil with stability in the short run, but at a very high price. The system gave new life to the old structures of power in the countryside. In effect, the modernizing, urban elites struck a deal with the traditional rural elites. The latter retained much of their power in the countryside in exchange for loyalty to the central government and the new "representative" republic. In the long run, the arrangement reinforced and perpetuated the worst features of the old power structure well into this century. Brazilian politicians constructed the façade of representative, democratic politics on the still-very-powerful structures of patronage, patrimonialism, hierarchy, and elitism.

The Brazilian state exercises its power and influence through an enormous and complex bureaucracy. Politicians come and go, regimes rise and fall, but the bureaucracy provides the nation with continuity. A survey in 1985 by the (rather ironically named) Ministry of Debureaucratization estimated that the state bureaucracy consisted of more than 60,000 agencies. The bureaucracy—federal, state, and municipal—employs about 6 million people, one of every six salaried employees in Brazil. The salaries of civil servants (known as public functionaries, or *funcionários públicos* in Portuguese) alone accounted for more than 60 percent of the federal budget in 1989.

While entrenched bureaucratic interests and career civil servants may stand in the way of new governments and administrations in all complex societies, each new Brazilian president has enormous potential to reshape the bureaucracy through the power of appointment. Brazilian presidents are able to appoint more than 50,000 public officials, including those in the most important policy-making positions. (By way of comparison, the president of the United States appoints a few thousand new officials.) Historically, putting people on the public payroll has been an important means for politicians to attract and maintain constituent support. Labor legislation makes it very difficult to fire civil servants, and this makes any effort to dismantle the bureaucracy problematic at best. The bureaucracy has long formed the sinews of power in the patrimonial state, and to attack it is to strike at the very nature of the system itself.

Since the colonial era, the bureaucracy has offered both civilians and military officers an important career path. Bureaucrats rise through the system by demonstrating merit, but also, more importantly, by developing influential connections (known as *pistolão*, "big pistol"). The ambitious must cultivate patrons who will protect them and promote them. Their fortunes normally rise and fall with those of their patrons. Some work their way through the bureaucracy pursuing technical careers, others move between political and bureaucratic careers (cabinet ministers are often a good example of this pattern). This personalistic system of clientelism provides a unique Brazilian style to what might at first glance appear to be a highly formalized bureaucratic hierarchy.

POLITICAL PARTIES

The power of the patrimonial state, the bureaucracy, and the elites has been reinforced over decades by the absence of strong and long-lasting political parties. In Western Europe and the United States, the formation of national political parties has created institutions that often serve as checks on the centralization of state and elite power. Political parties provide institutional means for resolving conflicts and act as mechanisms for distributing power horizontally rather than hierarchically. Historically, true national political parties in Brazil have been weak, unstable, or nonexistent.

Under the empire in the 1830s, loose coalitions emerged and split into Conservatives and Liberals, yet these "parties" did not have a national institutional presence, and despite proclaimed ideological differences, it was often difficult to distinguish a Conservative from a Liberal. (In Gabriel García Márquez' classic novel, *One Hundred Years of Solitude*, one character cynically remarks that the only difference between Liberals and Conservatives is the time of day they go to mass.) Very often, family networks or regional roots, rather than ideology, were the motivation behind one's choice of party label. The formation of the Republican Party (which advocated the end of the monarchy) took shape in the 1870s, but it did not even elect deputies to national office until 1885.

Under the less centralized First Republic, the Liberals and Conservatives disappeared (a sure sign of their institutional fragility) and state-level parties emerged, especially in the powerful states. The Republican Party in São Paulo, for example, dominated state politics, but like other state-level parties, did little in the way of institution-building other than getting the vote out for elections. The formation of the Brazilian Communist Party in 1922 (PCB for *Partido Comunista Brasileiro*) marked the appearance of the first truly national political party with an ideological program. Yet the PCB operated clandestinely for much of its early years, and it probably never attracted more than a few percent of the eligible voters. The Revolution of 1930 once again revealed the fragility of the party system, and Getúlio Vargas banned all political parties after the imposition of the New State in 1937.

The Vargas dictatorship temporarily ended a century of experience with poorly institutionalized parties and low levels of political participation. In the nineteenth century, fewer than 1 percent of the population voted in national elections. Throughout the years of the First Republic, from 2 to 5 percent of the population voted, and only three elections (1910, 1922, and 1930) were seriously contested. Not only were political parties unstable and weak, but they also did not stimulate participation.

Brazil's first real experience with national parties that appealed to the masses began in the mid-1940s, with the collapse of the New State more than a century after political independence. Three major national political parties emerged out of the transition from dictatorship to democracy: the Social Democratic Party (PSD), the Brazilian Labor

Party (PTB), and the National Democratic Union (UDN). These three parties dominated electoral politics from 1945 to 1964. The PCB hovered in the background, suffering from persecution and (after 1947) from government repression. A dozen or so minor parties appeared and played a role in coalitions during these two decades.

Both the PSD and the PTB had a more statist orientation than the UDN, which was dominated by Vargas's enemies and which had a right-wing ideological slant. Much of the PSD's strength came from the old political bosses, especially in rural areas, while the PTB appealed to the urban working class. For two decades, the PSD consistently controlled about one-third of the seats in congress, the UDN about one-fourth, and the PTB around one-fifth. When acting in a coalition, the PSD and the PTB could command a majority in the congress (which grew from approximately 300 members in the late forties to just over 400 in the early sixties).

Vargas played a role in the creation of both the PSD and the PTB, and the two parties dominated national politics from 1945 to 1964. PSD candidates won the presidency in 1945 and 1955. Vargas won with the backing of the PTB in 1950. Only Jânio Quadros, an outsider without a party but courted by all, broke the PSD-PTB presidential monopoly with his election in 1960. His vice-president, João Goulart, belonged to the PTB, and he became president when Quadros abruptly resigned in 1961. The UDN played the role of the frustrated outsider during these years. Although they backed Quadros in 1960, he was never a member of the UDN.

Vargas' rule from 1930 to 1945 moved Brazil from the traditional politics of the old-boy network to the mass politics of the postwar period. The military coup that deposed Vargas in 1945 unleashed the forces of the new Brazil and forced the elites to confront them. Contending groups within the elite scrambled to attract the votes of the masses as the number of eligible and participating voters mushroomed rapidly. In 1930, in the last presidential election of the "old regime," fewer than 2 million voters had cast their ballots (6 percent of a population of 33 million). In 1945, in the first free national election of the new era, nearly 6 million votes were cast (13 percent of a population of 45 million). By 1960, the number of voters in the presidential election approached 12 million (17 percent of a population of 71 million). In relative terms, the number of voters increased nearly 300 percent between 1930 and 1960. In absolute terms, the electorate increased 600 percent.

Open and unrestrained mass democracy, however, had not completely replaced the controlled elections arranged by a small political elite. The traditional elite (landowners, coffee planters, merchants) and the newer elite (industrialists and financiers) still wielded enormous power through their control of the economy, the media, and the state. They now used the instruments of power in an effort to manipulate the masses. The rules of the game may have changed, but the elites still held all the trump cards.

The most powerful challenge to the old-style politics came from populist politicians. Much like Huey Long and other populists in the United States, these politicians strove to mobilize the vote of the masses. Brazilian populists are charismatic figures who seek to forge a coalition that includes the middle and lower classes by stressing economic nationalism and social welfare programs. These politicians recognize that numerically the masses hold the key to winning elections, but that the middle classes are the key to success and stability in the political system. The task for populist politicians is twofold. First, they have to attract the votes of the masses with promises of the benefits of the welfare state and economic growth through protectionism. Second, they have to persuade the middle classes that economic nationalism offers the route to economic growth, and that the welfare state offers the way to social and political stability and peace.

Getúlio Vargas had instinctively understood this as early as the 1930s, and after 1945 he explicitly adopted populist tactics. Vargas and a host of other major political figures (Juscelino Kubitschek, João Goulart, and Adhemar de Barros, for example) dominated Brazilian politics from 1945 to 1964 through their populist appeals. Although they came from the traditional political elite, their populism broke with the tactics of traditional politics. The more conservative elites often saw the populists as revolutionaries. They were not. Populists did not fundamentally challenge the system. They attempted to rearrange the political equation, not to junk it.

In the twenty years following the end of World War II and the coup that deposed Vargas in 1945, growing political polarization plagued Brazilian politics. In the presidential elections of 1950, 1955, and 1960, populist politicians managed to mobilize the vote to defeat the candidates of the UDN and others who opposed the Vargas legacy. Yet once in office, the populists did not have the votes in congress or the political will to push for fundamental political reforms: voting rights for illiterates, land reform in the countryside, and civilian control of the armed forces. In a society in which half the population was illiterate, millions of peasants were landless, and the military regularly intervened in politics, these reforms would be basic to any hope of fundamental political change.

The two most skillful politicians of the era, Getúlio Vargas and Juscelino Kubitschek, tried to finesse the confrontation over basic reforms through economic growth. Both stressed nationalism and promoted domestic industrialization in a valiant effort to raise the standard of living for all Brazilians, rich and poor. Vargas reached a momentous political deadlock, and his response to the crisis was suicide. Kubitschek brilliantly improvised and offered something for everyone. He was a builder who presided over a period of phenomenal industrial expansion. In the end, however, he too could not escape the challenges of reform. Deficit spending, foreign debt, and inflation caught up with him at the end of his term, and he chose to pass the crisis on to his successor rather than impose an austerity program. Kubitschek found that he could give everyone a piece of the growing pie, but the cost of making the pie was too high.

Rather than establishing a stable and orderly democratic process in the postwar years, the new party politics constantly brought the political system to the verge of chaos. Vargas' suicide in 1954, the election of 1955, and the resignation of Quadros in 1961 each brought the system to the brink of collapse. On several other occasions, the threat of a military coup loomed near as the major political parties failed to establish a working consensus on the rules of the game. UDN power brokers, chaffing at their inability to win national elections, regularly urged sympathetic military officers to intervene in the political process. Throughout the fifties and early sixties, the fragile democratic system survived, despite continually heightened conflict among the major political parties. The major political parties contributed to the polarization of the early sixties, and the military coup in 1964 in some ways represented the failings of the postwar experiment with party politics.

Many of the new military rulers were ready to end party politics altogether, as military regimes in other Latin American countries would do in the sixties and seventies. This "politics of anti-politics" became a characteristic of dictatorships throughout the region. The moderate wing of the military regime believed that the party system could function once it had been "cleansed" of "undesirable elements." Through 1964 and 1965, thousands of politicians, bureaucrats, and military officers were stripped of their political rights by the regime which viewed them as dangerous. After the coup, the UDN enjoyed great influence in the government after years in the opposition. Leaders of the UDN had long called for military intervention and had played key roles in planning and provoking the coup.

The generals, hoping a purged system would produce strong support for the regime, allowed elections to proceed in 1964 and 1965. Much to their dismay, the PSD and PTB continued to win elections, and the UDN remained a minority party. After the PSD and PTB won 8 of 11 governorships in the October 1965 elections, the generals decided to construct their own party system. In October 1965, Institutional Act Number 2 (AI-2) abolished the old political parties and allowed just two new parties: the National Renovating Alliance (ARENA) and the Brazilian Democratic Movement (MDB). The ARENA became the official government party, and the MDB the only legal vehicle for the opposition. The generals wanted a form of party politics that they could control, and that would provide a stability lacking in the chaotic, multiparty system of the previous twenty years.

From 1965 to 1979, Brazil gained valuable experience with truly national political parties, but under a tightly controlled and artificially constructed system that severely limited party autonomy. The military regime centralized power and generally ignored the congress and both parties when making major decisions. In 1970 and 1972, ARENA dominated the national (congressional), state, and local elections. The government cancelled the direct elections for governors in 1974 and had the ARENA-dominated state legislatures elect governors.

Despite their efforts, the generals could not control elections indefinitely. Beginning in the mid-seventies, the MDB became an increasingly successful vehicle for opposing the regime. Although severely restricted and purged, it did provide the best legal means to oppose the military regime. In the mid-seventies, MDB electoral successes at the polls forced the regime to regroup and to rethink party politics. The government adopted a "divide-and-conquer" approach.

A new electoral law in 1979 allowed the formation of new parties under very restricted guidelines. The strategy of fragmenting the MDB initially worked as most members of the ARENA joined the new "government" Social Democratic Party (PDS), and the opposition split into four parties: the Party of the Brazilian Democratic Movement (PMDB), the new Brazilian Labor Party (PTB), the Workers' Party (PT), and the Popular Party (PP).

Since the early eighties, the return to civilian rule has been accompanied by the proliferation of political parties. Eleven parties entered candidates in the November 1989 presidential elections, and almost as many small parties combined forces with the eleven presidential candidates. In the late 1980s, with some 200 or so deputies (out of nearly 500), the PMDB (the successor of the old MDB) became the largest party in congress. Although the old MDB lost many of its former deputies to new parties, the PDS experienced an even more pronounced attrition. (So much for the military's divide-and-conquer strategy.) The PDS had about 100 deputies, which made it the second largest party. The Liberal Front Party (PFL), a breakaway faction of the PDS, had nearly 100 members in congress.

With the rapid proliferation of new parties, achieving a majority in congress has become increasingly difficult. After the PFL split away from the PDS in 1984, the PMDB became the majority party but continued to lose members and deputies with the formation of new parties, especially the Brazilian Social Democratic Party (PSDB). By 1990, no party could manage a congressional majority. In the October 1990 congressional elections, the PMDB remained the largest party but elected just 107 federal deputies. The PFL elected 87, the PSDB, 37, and the PDS, 43. Most of the remaining seats were divided between the PTB (36), the PDT (47), and the PT (35).

Brazilian politics looks more and more like a European parliamentary system, with many parties that must form coalitions to rule. Since the mid-eighties, Brazilians have debated whether to keep their presidential system or to adopt a European-style parliamentary system. A very small, but vocal, monarchist party also has managed to garner significant publicity for its attempts to restore the Braganza dynasty. Efforts to dump the presidential system failed in both the constitutional revision of 1988 and in a national plebiscite in 1993 (when Brazilians chose among presidential, parliamentary, and monarchical systems of government). Despite the efforts of influential politicians and intellectuals, Brazil appears likely to stick with its century-old presidential model.

KEY ACTORS AND INTEREST GROUPS

For at least a century, Brazilian politics has been the search for a consensus, a political formula that will foster stability, minimize class conflict, and promote economic development. The rise and fall of regimes—the cyclical swings from representative republic to dictatorship, from centralization to decentralization—testify to the failure of Brazilians to forge an lasting political solution to the conflicts and tensions in their society. For the past twenty years, as the country moved fitfully from military to civilian politics, social and economic crises have dramatized the need to find a workable political formula, and have also made the search for the formula more difficult. If Brazil does not find a political consensus soon, it may be condemned to remain the "country of the future."

Elites

The Brazilian elite are the key to understanding national politics and to establishing consensus. Historically, the elite (economic, political, bureaucratic, military, ecclesiastical) have dominated Brazilian politics. They have had (and continue to have) the power to make or break any political formula for the country. In the colony, metropolitan and colonial elites shared power uneasily. A small minority of sugar planters, gold barons, merchants, bureaucrats, priests, and military officers ruled over a huge majority of slaves, free blacks and mulattos, poor whites, and Indians. In the nineteenth century, the rise of coffee and later industry diversified the elite both economically and regionally. They continued to dominate politics, but faced increasing challenges from below, especially in the so-called regional revolts and uprisings that frequently shook the fragile façade of social peace.

With the diversification of the economy during the last hundred years, the power of the land-owning elite has diminished, although they remain a powerful force. Industrialists, bankers, and business executives have emerged as forces within the elite. As their composition has become more diversified, the elite's ability to forge a unified front has also become more complicated. Although the land-owning elite of the nineteenth century were not a monolithic block, they certainly shared more economic and ideological baggage than today's elite. They were fewer in number, male, and generally descended from landed families; they also attended the same schools and rose through the ranks of the Liberal and Conservative parties. The contemporary elite also must fight to maintain their political power through the rules of modern representative democracy; they cannot rely on the exclusionary (and openly repressive) politics of nineteenth-century constitutional monarchy.

But who are these elusive elite? Traditionally, the social and economic elite have been composed of large landowners, powerful merchants, wealthy professionals, high-ranking military officers, and influential clergy (generally bishops and archbishops). Since the end of the nineteenth century, and especially since the 1930s, industrialists, executives in multinational and state corporations, and financiers have entered into the elite ranks. These simple labels, however, do not do justice to the complexity of the elite. Often, one individual may simultaneously be a landowner, banker, and industrialist, which makes glib generalizations about different interest groups complicated and dangerous.

Wealth, family connections, and the ability to shape events define members of the elite. Generally, although not always, membership in the elite requires substantial financial resources. Using a crude economic indicator, some 1 percent of the population would probably be eligible. (The top 10 percent of the Brazilian population accounts for nearly 50 percent of all individual income. The bottom 70 percent earn about 25 percent.) The societal elites often have excellent family connections and wealth, although family history sometimes overrides (at least temporarily) declining resources. What truly narrows the numbers of elites is the ability to shape events. The elite are the decisionmakers (politicians, planners, industrialists, bankers, generals) whose choices and actions shape the lives of the vast majority of Brazilians. When one cross-references all three factors, perhaps less than a million people compose the elite, some tens of thousands exert most of the power, and a few thousand (in a society of 150 million people) stand at the top of the pyramid of power and influence.

The elites constantly change and reinvent themselves. Although some can trace their roots back to the colonial era, even more come from families that achieved their wealth during the nineteenth century. Many, such as the Matarazzo, Guinle, and Klabin families, trace their fortunes back to immigrants who arrived in the late nineteenth or early twentieth centuries. As in many Western societies, the *nouveaux riches* immigrants have intermarried with socially prominent families with long-standing elite credentials. The explosive economic success of São Paulo in this century has produced a large portion of the current economic elite. Waves of immigrants from Europe and the Middle East have also transformed the "Brazilian" elite of the nineteenth century.

Historically, Salvador, Rio de Janeiro, and São Paulo have been the principal centers of elite power and influence. Although Salvador remains an important regional center, its status fell markedly after the decline of sugar and the rise of gold mining in the eighteenth century. From the early nineteenth century until recent years, Rio was the undisputed elite center since it was the economic, political, and cultural capital of the nation. The rise of São Paulo after 1920 as the nation's economic capital, the transfer of the national capital to Brasília in 1960, and the growing cultural sophistication of São Paulo since the 1940s have made the *paulista* capital at least as important an elite power center as Rio (and debatably more so).

The regional elite in the various state capitals also form an important part of the "ruling class" in Brazil. As the nation has become more unified, and the power of the state has extended into all parts of the country, the regional elite have become integrally tied into a national power network. Very often the state-level elite (especially in the Northeast) have been dominated by clans vying for power. In contemporary Brazilian politics, these historic family ties now play a role on the national scene. Several generations of the Magalhães family, for example, have played powerful roles in Bahian politics. In recent decades, Antônio Carlos Magalhães, three-time governor of the state and the son of a former federal deputy, has been a major player in national politics. His son, Luís Eduardo, has now become a prominent figure in the PFL and was elected as president of the chamber of deputies in January 1995.

Perhaps the most striking feature of Brazilian politics is the ability of this constantly adapting elite to dominate the political process over five centuries and in a variety of regimes, from monarchy to republic to dictatorship. Despite the massive economic, social, and political changes in Brazil during the last 150 years, a relatively small elite continues to dominate politics. Their task has become increasingly complex and difficult since the turn of the century, and especially since 1945, with the rise of new social groups such as the urban working class and middle class. Even in the era of mass politics, they have managed to survive and manipulate the political process to their advantage. Into the 1920s, they maintained power through the manipulation of the electoral process, thus creating a "controlled" representative republic. In the New State and during military rule, they formed coalitions among themselves and with the military to share power and keep the masses at bay. Even in the two decades of postwar populism, and despite serious intra-elite divisions, they managed to hold onto power. Politics, and populism in particular, in the postwar years has been clientelism on a mass scale. Elites accumulate and dispense resources to control the political process. The persistence of elite power in a society with more than 90 million voters has led the Argentine political scientist Guillermo O'Donnell to label Brazil a "delegated democracy." The masses continue the centuries-old Iberian tradition of relinquishing power to the state and the elites.

Military officers form a special kind of elite within Brazilian politics. Often lacking the wealth and family connections of the economic and social elite, military officers achieve elite status through their ability to shape politics. Since the overthrow of the empire in 1889, the Brazilian military has frequently and forcefully acted as the ultimate arbiter in national politics. In the five years following the coup, Brazil became a virtual military republic, and for the next thirty-six years the military guaranteed the stability of the "politics of the governors." Military dissatisfaction and elite divisions led to the bloody Revolution of 1930. In 1937, Vargas imposed his New State upon the populace with the collaboration of the military, and he was deposed in 1945 when the military high command decided it was time for him to go. Throughout the 1950s, the military continually intervened in national politics to guarantee elections and the transition of presidents.

The 1964 coup definitively marked the culmination of the rise of the military as *the* power broker in Brazilian politics. After seventy-five years of cultivating an increasingly powerful role as arbiter through intervention in major crises, the military now fully asserted itself and took power. Rather than arbiters, the generals became the dictators of

Brazilian politics. After decades of increasing frustration with the nation's civilian leadership, the military decided to show the politicians how to run the country. As in previous interventions, the generals acted in conjunction with other sectors of the elite. What was surprising (especially to these elite) was the military's resolve to stay in power this time rather than to act, and then step down.

Despite the widespread image of all Latin American military officers being stamped out of the same mold, the Brazilian military is not—and has never been—a monolithic institution. The officer corps has long been a cross-section of the middle sectors of society, and has reflected the ideological diversity of the middle class. In this century, Marxism, fascism, populism, and representative democracy (to name just a few positions) have all had their supporters among the officer corps. Factionalism and internal struggles pervade the military, just as they do any large bureaucratic organization. The key to the military's success has been its ability to impose an internal unity at crucial moments. Unity has allowed the military to act decisively and to achieve its political objectives. When that unity has been lacking, the military has refrained from intervening in politics, or has faced the possibility of internal fragmentation and institutional breakdown.

Education and training have been the most powerful force in molding a unified military and in the creation of a military ethos. Most high-ranking officers have followed the same career path. Army officers, for example, graduate from military high schools, attend military academies, and (if successful) go on to the Army General Staff College (ECEME) and (for the cream of the crop) the Higher War College (ESG). As a rule, an officer's class rank—and the network of friends he cultivates at the schools—determines the speed and number of promotions through the ranks. (This is very different from the U.S. Army, in which officers are drawn from academies, R.O.T.C. programs, and the enlistment process, and in which combat experiences create a more diverse and multi-track officer corps.) Since all high-ranking Brazilian officers have been through the same schools and training programs, they share very similar experiences and career paths. They may represent various regions, have different family backgrounds, and hold distinct political views, but they all have punched the same tickets.

The coup in 1964 represents the triumph of a mentality that had been developing within the military, particularly the army, for nearly a century. In many ways, the military regime became the expression of the positivist dream of the late nineteenth century. The late-nineteenth-century positivist ideologue, Benjamim Constant, and the young officers he influenced in the military academy, wanted to see a Brazil that incorporated the latest advances in science, technology, and industry. They believed that an authoritarian republic run by "experts" was the best way to create this new Brazil, the country of the future. The positivists briefly had their taste of power when they helped create the First Republic. They even left their motto, "Order and Progress," stamped on the Brazilian flag.

In succeeding decades, the positivist army officers and their intellectual heirs grew increasingly frustrated with the civilian political leadership. The *tenente* revolts of the twenties, the civil war in 1930, and the periodic military interventions in national politics reflect the military's frustrations with civilian leadership. Although the officers who created and guided the military regime after 1964 were not positivists, and would probably deny any direct links to the positivists, their desire to create an industrializing, technologically sophisticated Brazil with an authoritarian republic run by technocrats makes them the ultimate heirs of the positivist mentality. They created the Brazil that the positivists had envisioned at the turn of the century.

A well-refined vision known as National Security Doctrine (NSD) guided the military's plan of action in the 1960s and 1970s. NSD had its origins in the military schools as an offshoot of the Cold War. In 1949, the Brazilian military created a Higher War College (*Escola Superior de Guerra,* or ESG) that was modeled after the U.S. Army War College in Carlisle, Pennsylvania. Many of the key officers in the ESG were veterans of the Brazilian Expeditionary Force (FEB) that fought alongside General Mark Clark's U.S. Fifth Army in Italy in 1944-45. (Brazil was the only Latin American country to send combat troops to fight the Axis during World War II.) Deeply influenced by training in the United States, and by the work of U.S. instructors in Brazil, officers developed a sophisticated view of how their country fit into the superpower struggle. In essence, they saw the world divided between two contending camps, one Christian and democratic, the other atheist and communist. Third World nations stood on the frontlines of the conflict between these two camps, which were led by the United States and the Soviet Union. According to the logic of NSD, these two camps were engaged in a world war to destabilize their enemies' allies through internal subversion. This was not a conventional war, but an unconventional struggle of subversives attempting to spark leftist revolutions against the allies of the United States.

The Brazilian generals watched the turmoil brewing in their country during the early sixties and saw another Vietnam—or, more precisely, another Cuba—in the making. Their political objective was to halt the drift toward revolution, and to completely destroy any forms of internal subversion. Once "order" had been restored and revolution repulsed, the military could then embark upon a program of nation-building to make Brazil a power in the world community. The generals understood that an industrial economy held the key to power in the international economic and political arena. "Progress" meant economic growth, industrialization, and technological sophistication. "Security and Development" became the military's updated version of "order and progress."

After 1964, the military became the elite of the elite. They could not rule without support from the other elites, who could not rule without the military's backing. As long as repression and economic growth kept the masses in check, and kept significant sectors of the elite relatively satisfied, the military had its way. Severe repression conclusively annihilated any threat of leftist challenge by the mid-seventies, and the oil shocks and escalating debt began to unravel the economic "miracle" by the end of the decade. In the 1980s, major actors within the economic elite—as well as the mobilization of workers, the middle class, and the church—weakened the regime and forced the generals to retreat to the barracks. Chastened, but not defeated, the military pulled back from direct rule while other sectors of the elite and of society once again attempted to reach a national political consensus.

The Middle Class

Historically, military officers have come out of the ranks of the middle class, that most elusive of all actors in the political process. While social scientists debate whether or not a "true" middle class exists in Latin America, I will use the term loosely to refer to those in Brazil who are neither from the elite, nor from the poor and working classes. This is (admittedly) a vague and amorphous group of people, sandwiched between a very powerful minority and the largely powerless majority. By Brazilian standards, middle class people have comfortable lives with monthly incomes above $1,000 ($12,000 in annual income, or ten times the minimum wage). They usually own a home or apartment, have at least one car, and send their children to private schools and ultimately to a university. These are the civil servants, bureaucrats, and professionals in Brazilian society. They are not rich, but (by income standards) they are in the top 6 percent of the nation. (Another 24 percent of income earners—the working class—fall between the middle class and the 60 percent of Brazilians who live in poverty.)

Part of the problem with defining the middle class stems from the historical differences between them and their counterparts in the North Atlantic world. The middle class emerged as an affluent, vocal, and politically potent force in Europe and the United States during the nineteenth century. By the twentieth century, they had effectively gained control of politics and become the majority in the developed world. Two-thirds of U.S. families fall into the middle class when they are defined using an income measure (roughly a $20,000-60,000 annual family income).

The Brazilian middle class, on the other hand, has been unable to achieve power or a majority in the political system. Uncomfortably situated below the elite—with whom they share in the economic growth of the past century—and the poor majority, the middle class has flirted with reform and has embraced dictatorship. Throughout Latin America, the middle class has been the principal force behind many reformist political parties, such as the Christian Democrats in El Salvador or Chile, and the National Liberation Party in Costa Rica. In the 1960s, many U.S. and European observers and governments saw the middle class as the vehicle for change, the means to break down the stranglehold of old land-owning elites over the peasant masses. The U.S. Alliance for Progress explicitly sought to build up the middle class and their reformist parties in order to open up the Latin American economies and political systems.

Unfortunately, when the reformist, middle-class parties pressed the elite for true change (read: a share of political power) the elite in many places became more intransigent. The rise of armed leftist revolutionaries reinforced the attitudes of the elite, and frightened the middle class. In country after country—Argentina, Brazil, Chile (the list goes on)—pressures for reform often accompanied or provoked more radical movements. The political crises of the 1960s and 1970s brought many countries to the verge of chaos or civil war, and the military then took control.

Brazil in 1964 was the classic example. Although no single party represented the middle-class reformers, several parties (especially the PTB and PSD) contained strong reformist wings with middle-class support. Yet, when the country approached gridlock in March 1964, the middle class became frightened by the specter of leftist revolution, and political and economic chaos. As in Argentina (1966) and Chile (1973), when the going got tough, much of the middle class moved to the right and supported military intervention. The "middle-class military coup" (in the words of Argentine political scientist José Nun) swept across Latin America between 1964 and 1973. The middle class feared revolution and instability more than the elite's continued domination of the economic system and political process.

The experience of the sixties and seventies raised tough questions for those of us who would like to see more democracy and less inequity in Latin America. How can true (relatively peaceful) reform be achieved if this articulate, educated, and highly organized middle class fails in its efforts? Put another way, how can the Brazilian elite be persuaded and/or pressured to open up the political process and make social reforms as long as they keep the masses in their place and refuse to respond to the efforts of the middle class through traditional political channels? In other words, if the middle class cannot get the elite to ease up and open up peacefully, who can?

With the redemocratization of Brazil in the 1980s, the middle class has again asserted itself by organizing and joining political parties. As in the postwar years, party proliferation has diffused the impact of this class. The PSDB, PMDB, and PFL (to name just a few examples) all draw significant middle-class support. They represent reformism—from left to right of center—but spread across political parties. The problem of the middle class is the problem facing Brazilian politics: how to forge a coalition of forces for change? The elite (including the military) continue to dominate politics with the acquiescence (even if reluctant) of the middle class. The only way to open up the political process and create social mobility is for significant sectors of the middle class to forge political coalitions with parties that represent the interests of the poor and working masses. Yet, they must do this gradually and in a way that does not set off another round of confrontation and crisis. In short, they cannot give the elite an excuse to (once again) invoke military intervention.

The Working Class and the Working Poor

Rural and urban workers have long suffered from repression and manipulation by the elite, the military, and the middle class. Until this century, poor peasants made up the majority of Brazilians but were a small voice in the political process. Except in times of crisis and regime breakdown, the rural masses rarely made their voices heard. Many of the regional rebellions and social uprisings of the nineteenth century forced the elite to take notice, but the central government ultimately crushed all revolts and the "inarticulate" masses had little or no institutionalized political representation. Through persuasion or intimidation, the rural colonels controlled and manipulated the poor, illiterate masses who could not organize.

With the urbanization and industrialization after 1900, many of these peasants migrated to the cities in search of a better life. The growth of an urban working class began to send tremors through the political system as early as World War I, and by the 1930s Getúlio Vargas and others had begun to recognize the need to harness this new force. Ever the conservative modernizer, Vargas sought to "make the revolution before the people do." Unlike in Europe and the United States, the urban working class in Brazil (much like the middle class) was small and politically weak. They became a force, but not a dominant one. Often manipulated in the populist politics of the 1940s and 1950s, the urban working class gained an institutionalized voice in the political process, but largely through government-dominated unions and a political party (the PTB) controlled by populist politicians like João Goulart. Under military rule, the government "intervened" into unions, expelling suspected subversives and making sure that union leaders were, at a minimum, politically timid and cooperative.

In the 1970s, a new generation of union leaders, best symbolized by Luis Inácio (Lula) da Silva, emerged and began to construct a dynamic working-class movement. This movement, which began on shop floors during the process of redemocratization, eventually became a broad-based social movement that demanded full citizenship rights and social legislation for the masses. In stark contrast to the situation in the developed world, union membership rose dramatically in Brazil during the 1980s—from 12 to 30 percent of the workforce. In 1980, the Workers' Party (PT) became the institutionalized political vehicle for this movement. Lula became the party's leader and its presidential candidate in 1989 and 1994.

Ironically, the organizational and political success of organized labor, in effect, moved it closer to the middle class in economic terms. Lula's *paulista* steelworkers, for example, succeeded in gaining wages and benefits that moved them away from the vast majority of the urban masses and closer to the living standard and lifestyle of the middle class. Today, three-quarters of the steelworkers earn at least five times the minimum monthly salary, or about $500 a month or better. (About 90 percent of all Brazilians earn less than that.) More than 90 percent of the homes in the heavily working-class suburbs of São Paulo (the so-called ABC region) have color television sets, and more than 70 percent have a washing machine. These figures are substantially higher than those for cities such as Curitiba, Salvador, Porto Alegre, and Recife.

The 70 percent of the urban working class who are not unionized and who do not benefit from social and labor legislation, form an underclass with very little political voice or power. Along with the rural poor, they are at the very bottom of the political power pyramid in contemporary Brazil. For these people, urban poverty is recycled rural despair. They form Brazil's "silent majority." Because they have the lowest levels of education and organization, and the barest of incomes, the political maneuvers of the elites, the military, middle class, and organized labor largely shape the fate of these rural and urban poor people.

The Catholic Church

In the past few decades, the Catholic Church has emerged as a vibrant force on the side of the oppressed rural and urban masses, and as a powerful moral voice in support of human rights. The politicization of the Catholic Church and its emergence as an activist force paralleled the emergence of the military in the 1950s and 1960s. While the military developed National Security Doctrine as its worldview, the Catholic Church was developing its own vision of Brazil.

Paradoxically, the Catholic Church in Brazil has become more visible as a political and social force over the last century as the country has become more secular, more modern, and less Catholic. The revival of the church began, strangely enough, with the fall of the empire (which had linked church and state), and the proclamation of the republic (which officially separated church and state). The secular republic freed the church from the institutional strangulation it experienced under Pedro II. For the first time in decades, church officials could expand their administrative apparatus without the direct obstruction of the government. The separation of church and state by the liberal republicans gave the Catholic Church new life.

Under the influence of Catholic social action ideology and programs in Europe, the Brazilian Catholic Church worked diligently to organize the faithful into politically active organizations. To combat the rising influence of socialism and communism among urban workers, the church promoted the formation of Catholic labor organizations. The church also encouraged the formation of Catholic organizations to put the church back into an active social and political role in the community. Catholic social action espoused a "third way," rejecting the godless atheism of socialism and communism, and the individualism and materialism of capitalism. With collectivist and communitarian values, Catholic social action had much in common with the socialist ideologies it rejected. Yet the movement strongly supported private property and capitalism, albeit with strong regulation to produce what has often been called "capitalism with a human face."

The Catholic social action movement emerged in the 1920s and 1930s in Brazil, paralleling a trend in the rest of Latin American and predominantly Catholic European nations. These years marked the reemergence of the Brazilian Catholic Church as a political force, albeit a weak one. Over the next few decades, Catholic clergy and laity would become increasingly politically active—and ideologically divided.

The rise of liberation theology in the 1950s and 1960s would divide the Catholic Church in Brazil and in the rest of Latin America. The movement known loosely as liberation theology began to emerge in the 1950s as priests increasingly found their mission among the poor rather than the affluent. Liberation theology exploded onto the scene in Latin America in the 1960s in the aftermath of Vatican II

(1961-63) and the Latin American bishops' conference at Medellín, Colombia, in 1968.

Both events solidified the fundamental features of the movement: an emphasis on working with the poor (a "preferential option for the poor"), and the combination of a message of worldly redemption with the traditional message of spiritual salvation. Many in the Latin American Catholic Church in the fifties and sixties argued that the church was not the institution and the hierarchy, but the faithful. In Latin America, that meant the huge numbers of impoverished people.

Breaking with the traditional message of the Latin American church, activist priests began to emphasize Jesus, the revolutionary figure of the Sermon on the Mount, the Christ of the poor, oppressed, and downtrodden. ("The Spirit of the Lord *is* upon me, because he hath anointed me to preach the gospel to the poor; he hath sent me to heal the brokenhearted, to preach deliverance to the captives, and recovering of sight to the blind, to set at liberty them that are bruised. . . ." [Luke 4:18] Rather than telling the poor masses to accept their suffering and the status quo, and to await their reward in heaven, the activists worked to help the masses change their lives in this world. They argued that salvation or liberation was not simply spiritual, but also temporal. In a message eerily reminiscent of Calvin and the Protestant Reformation, these priests argued that to be a good Christian one had to behave like a good Christian. Belief required action. The effort to carry out Jesus's message that "the meek shall inherit the earth" would be radical in any society. In the context of Latin America, with its enormous amount of poverty, the message is revolutionary.

Beginning in the 1950s, and accelerating in the 1960s, activist priests organized what became known as Christian Base Communities (or CEBs, the acronym for *comunidades eclesiais de base* in Portuguese) to bring the faithful together to study and work together in a Christian environment. This work had both spiritual and practical aims. Priests have always been in short supply in Brazil (as in most of Latin America), and one thrust of the pastoral work with these communities was to delegate some of the traditional responsibilities of the clergy to lay persons. Brazil has just 1 priest for every 9,000 Catholics. From a practical standpoint, responsibilities had to be delegated to the laity if the church was going to restore and maintain an active presence among the masses. The church needed to empower the laity in a ritual sense, or risk losing them.

Inevitably, this work also carried a political message: the poor do not have to accept powerlessness. Activist priests, nuns, and layworkers sought to empower the masses through the creation of CEBs. Committed Christians in the CEBs worked to understand biblical lessons, and to translate those lessons into the transformation of their daily lives. These communities focused on giving the poor access to land and housing, and helping them develop basic services—water, health care, and education. Along with Central America, Brazil has been one of the most important foci of the development and expansion of CEBs. Today there are (by some estimates) more than 80,000 base communities in Brazil that reach hundreds of thousands of people. These communities, organizations from below, began to pressure the traditional hierarchy to address issues at the very core of economic and political power in Latin America.

This movement from below also split the episcopal hierarchy of the largest Catholic country in the world. Increasingly, the more than 350 Brazilian bishops divided between those who supported the movement and those who saw CEBs and liberation theology as nothing more than marxism disguised as Scripture. The National Conference of Bishops of Brazil (known as CNBB for its Portuguese initials), the third largest national conference in the Catholic world (after Italy and the United States), divided between the activists, traditional conservatives, and moderates. Throughout most of the 1970s and 1980s, the CNBB divided almost equally among the three wings. The activists attracted enough support from the moderates to make the Brazilian Catholic Church one of the most theologically progressive and pastorally active in the world.

Helder Câmara, archbishop of Recife, became the most famous and eloquent spokesman of the new theology among the church hierarchy. In his sermons, speeches, and writings, Câmara appealed for recognition of the concerns of the Brazilian masses. After the military coup in 1964, Câmara became the most prominent religious critic of the regime. He led the majority of the Brazilian bishops in a sustained attack on the repressive nature of the regime. With virtually all critics crushed by censorship and torture, the church became one of the few vehicles (and perhaps the most important) for speaking out against the repression.

The church paid a high price for its courage. A number of activist priests were arrested and tortured. Several were assassinated by right-wing death squads. Clerical garb no longer served as a shield. Activist priests and nuns now became targets of the repressive military regime along with leftists, labor leaders, and those who spoke out against the regime. The military regime simply deported priests of foreign nationalities who dared to speak out against the repression.

Paulo Evaristo Arns became the most visible representative of the progressive wing of the church hierarchy in the 1970s and 1980s. In 1970, he was appointed archbishop of the city of São Paulo (at that time the largest Catholic diocese in the world), and he later was named a cardinal. An outspoken critic of the military regime, Arns directed the work of the archdiocese toward the urban poor. Beginning in 1979, a group from the archdiocese secretly copied and analyzed the trial proceedings of Brazilian military courts to produce a very authoritative tally of the price of military repression. When the results of their study were published in 1985, *Brasil: Nunca Mais* (translated into English as *Torture in Brazil*) rapidly sold more than 200,000 copies and stayed atop the bestseller list for six months. (The typical press run for a book in Brazil is 3,000 to 5,000 copies.) According to the report, 333 people died from torture, assassination, or "disappeared" between 1964 and 1981. One-fifth had died in a failed rural guerrilla insurgency in the eastern Amazon in the early seventies.

The military's use of torture, and the Brazilians' reaction to the carefully documented repression, tell us a great deal about Brazilian politics and society. Although the repression in Brazil was very real, and very intense, it pales in comparison to the military regimes in Uruguay, Argentina, and Chile in the 1970s. An official Argentine commission has verified nearly 9,000 murders and "disappearances," and the number could be almost twice that. Even using the lower figure, the per-capita toll in Argentina (2,647) is nearly 100 times that of Brazil (279,279). Even in times of its most ferocious and bitter political conflict, the Brazilian military regime did not come close to the levels of political violence witnessed in the nations of the Southern Cone.

Cardinal Arns also opened his churches to the labor movement in São Paulo in the late seventies and early eighties. Banned from holding rallies, the workers would meet in cathedrals, stadiums, or plazas, ostensibly to celebrate mass, but in reality to demonstrate the huge support for striking workers. Arns and the church, in effect, openly challenged the government's efforts to repress the workers' movement. Although the Catholic Church never took an "official" political position in elections or in support of specific parties, large numbers of priests, nuns, and lay workers openly supported the Workers' Party (PT).

The emergence of the PT and the political activism of the Catholic Church underscore the shifting alliances and coalitions among the interest groups and actors in Brazilian politics during the last thirty years. The age of landowners, merchants, and peasants disappeared long ago, giving way to a more complex social and political puzzle. At least since the 1930s, Brazilians have struggled to put the puzzle together, even as new groups and actors proliferate. At all social levels, Brazil has become more complex, diverse, and nuanced. This complexity contributed to the political conflicts of the 1950s and 1960s. The generals tried to impose a political solution that would bring the highly diverse political and economic elite into an alliance with the equally varied middle class, and to prevent upheaval by the mushrooming urban lower class and the still-sizable rural masses. The military regime decided to use force to bring order to the puzzle.

CONTEMPORARY POLITICS

The guiding principles, the mechanisms of power, and the interest groups I have sketched in the preceding pages provide the keys to understanding Brazilian politics since 1964. Elitism, hierarchy, centralism, and patrimonialism (albeit in new forms) continue to play a powerful role in Brazilian political culture. The state and the bureaucracy remain at the core of the political system, while political parties strive to establish themselves as countervailing forces in the system. The elites dominate a democratic, representative regime in an uneasy relationship with the middle class and the most organized sectors of the urban working class. "People in search of a political voice" perhaps best characterizes the urban and rural poor, who remain weak.

Contemporary Brazilian politics is the search to construct a new order out of the legacy of the military regime. The armed forces briefly imposed a new political order out of the 1964 coup, and then watched it disintegrate in the 1980s. During the past fifteen years, civilian politicians have struggled to pick up the pieces and to create a stable and enduring order built not on repression and dictatorship, but on negotiation and elections. Brazilians have striven to extend the fruits of full political citizenship to everyone, regardless of class or personal connections. They have fought make their politicians accountable. The results have been mixed.

The events of 1964 reveal all the fissures and cleavages in the political landscape of contemporary Brazil. By March 1964, the polarization of right and left had reached unprecedented levels. The supporters and enemies of Vargas's legacy bitterly confronted each other. Nationalism further fragmented the nation along lines that were not clearly right nor left. The traditional elite in national politics continued their efforts to hold onto power while confronting the growing power of the new elite as well as the increasingly mobilized masses. The crisis in 1964 dramatically exposed the frailty and fragility of democratic institutions and processes despite nearly 150 years of experience with some form of representative government.

The generals (and a few admirals) staged their coup in 1964 with two major goals: to halt the perceived threat of a leftist revolution, and to revitalize an economy in shambles. As they saw it, economic and political mismanagement had brought the country to the brink of social chaos. Although the left became the main target of their repression, they surprised politicians on the right and center by taking aim at them as well. The military firmly believed that *all* of the civilian political leadership shared the blame for the crisis. The left might be the instigators, but their opponents had acted irresponsibly by not meeting the challenge of the left through the constitutional process. The military surprised their civilian allies by taking control of the system in 1964 rather than simply stepping in to eradicate the left. Pursuing what some have labelled the "politics of anti-politics," the generals chose to show the civilians the errors of their ways. Seventy years after Floriano Peixoto handed over the presidency to civilians in the First Republic, Brazil once again became a military republic.

The military regime (1964-85) reshaped some deeply rooted political traditions. With sophisticated technology and technocrats, the military centralized power and reinforced hierarchical structures with a vengeance. Elite control of politics took on a new meaning as the military became the dominant elite interest group. Social control, especially of the masses, now became more than an objective of the upper classes; it became a crucial pillar of national security. Finally, despite the military's expressed dissatisfaction with the corruption imbedded in the old politics of patronage, the generals would eventually bring disgrace on themselves with a patronage spoils system of unprecedented dimensions.

During the years of military repression, one political party provided the only possible "legal" voice for the opponents of the regime: the Brazilian Democratic Movement (or MDB). Those in the political system who had survived the purges and repression, and who still hoped to confront the regime through "official" and "constitutional" channels, did so primarily through the MDB. It brought together politicians of very diverse ideological perspectives who were unified in their opposition to the military regime. For two decades, the MDB made use of its limited rights to press the generals back toward civilian rule. Once again, the central state weakened and manipulated the party system, preventing its evolution into a full-fledged institutional force in the political process.

In the late sixties and early seventies, during the Costa e Silva and Médici administrations, the hopes of change through legislative channels and through the "official" opposition party seemed bleak. With the inauguration of Ernesto Geisel in 1974, however, the long, slow process of political opening (known in Brazil first as *distensão* and then later as *abertura*) had begun. As Alfred Stepan has pointed out, this was a process of liberalization initiated by the regime in a period when there was no significant political opposition exerting pressures and no major economic crisis, and when the coercive apparatus was completely intact. Over the next fifteen, years Brazil inched back toward civilian rule in what surely must be the longest process of "redemocratization" in the twentieth century.

This extended process vividly illustrates some of the enduring and essential features of Brazilian politics. As the masses began to mobilize, contending elite groups negotiated and bargained. Geisel and his key advisor, General Golbery do Couto e Silva, devised a series of measures to ease the return to civilian rule while attempting to prevent the return to the "excesses" of the early sixties. Their principal tactic was to maintain the official government party's majority in the Senate and House, as well as its control of state governorships, while fragmenting the opposition movement. Under the presidency of João Figueiredo

(1979-85), in an effort to weaken the MDB, the government implemented a new electoral law that led to the creation of a number of
opposition parties. Despite the Machiavellian maneuvers of Geisel and
Golbery, each time the regime allowed open and competitive elections,
the opposition scored major successes.

As the economy deteriorated in the late seventies and early eighties,
the business elite began to pressure the regime, both publicly and
privately, to relinquish power. The major task facing Figueiredo was to
conciliate between the contending political interest groups and to ease
the regime back to civilian control. The mobilization of the masses
increased the pressure on the political elite to make the transition. In
1978 and 1979, the long-dormant labor movement reawakened, and the
steelworkers in Greater São Paulo defied the regime with bitter strikes.
With the support of the progressive Archbishop Arns, they held mass
rallies in the cathedrals of São Paulo in open defiance of the government.

With the masses mobilizing, the economy deteriorating, and the
traditional political elite pressuring for a change, the military found itself
under siege. Figueiredo turned out to be an inept negotiator and
politician. By 1984, the generals had lost control of the process of
abertura. In the largest mass demonstrations in Brazilian history,
millions of people flocked to huge rallies all over the country and
demanded direct presidential elections. While tens of thousands of
Brazilians surrounded the congress in Brasília chanting *"diretas já"*
(direct elections right now), the congress narrowly voted down a law that
would have allowed direct elections in 1985.

The campaign for direct elections demonstrated both how much
and how little Brazilian politics had changed. The mobilization of the
masses across the country, especially around a single cause and in a
challenge to the military regime, clearly had no precedent. The campaign signalled the return to mass politics on a scale unknown since the
early 1960s. Yet, in the end, the movement failed. Despite the intense
pressure from below, the powers from above in the government wheeled
and dealed and bargained to defeat the proposed law. In an era of mass
politics, the elite continued to dominate the system. Compromise and
conciliation, especially among contending elite groups, remain the keys
to understanding Brazilian politics.

President Figueiredo bungled the indirect election in the "stacked"
electoral college and played right into the hands of the opposition.

Hoping to appease those who wanted to see an end to military rule, he set in motion the process to nominate a "safe" civilian for the presidency. Paulo Maluf, the former governor of São Paulo, emerged as the leading contender for the nomination. A wealthy businessman of Lebanese origins, Maluf had a knack for alienating powerful figures, both in the opposition and the PDS. Figueiredo ultimately refused to support any candidate publicly, thus acknowledging that he had lost control of the electoral process.

In the meantime, the PMDB (the new name for the old MDB) had managed to unify the opposition parties around the candidacy of Tancredo Neves, the highly respected governor of Minas Gerais. (As so often in twentieth-century Brazil, *paulista* and *mineiro* politicians once again dominated the struggle for the presidency.) Neves had begun his political career as a protégé of Vargas in the early fifties, had served as prime minister during the shortlived experiment with parliamentary politics in the early sixties, and had been a leading voice in the MDB during two decades of military rule. A grand-fatherly figure known for his ability to broker political deals, Neves was perhaps the only figure who could unify the opposition and pacify those in the government party who were nervous about the return to civilian rule.

As the vote in the electoral college approached in late 1984, Neves forged a broad coalition (known as the Democratic Alliance) with other opposition parties and a breakaway group from the ruling party. The deal he brokered with a dissident wing of the PDS snatched certain victory from Maluf and the generals. A group in the government party who rejected the Maluf candidacy broke from the PDS and called themselves the Liberal Front. Led by Aureliano Chaves, Figueiredo's own vice-president and a presidential candidate himself, and José Sarney, the leader of the PDS in the senate, the Liberal Front joined the opposition. (They also formed the core of what became the PFL or Liberal Front Party.) In classic fashion, Neves bargained with the Front and the leaders of the opposition parties to divide the anticipated spoils of victory. The PMDB and its allies drew up detailed lists of key government posts and assigned parties to them. As almost an afterthought, Sarney became Neves's vice-presidential candidate, thus symbolizing the deal between the Front and the opposition. Political regimes might come and go, but patronage lived on.

In the end, political brokers in backrooms rather than the masses in the streets made possible the transition from military to civilian rule. As Alfred Stepan has argued, the generals lost control of the political process they had initiated in 1964, and very astute civilian politicians then helped the disgraced generals make their exit. Even before the vote of the electoral college in January 1985, the results had been determined by the deals cut by Tancredo Neves and the power brokers in Brazilian politics. The scheduled inauguration of Neves as president on March 15, 1985, was to mark the penultimate step in a process of *abertura* that had taken a decade and that would not be complete until Brazilians could go the polls and freely elect their own president.

As the date for the inauguration approached, the victorious opposition parties spoke almost euphorically about a "New Republic" and a new era in Brazilian politics. The nation seemed to be entering uncharted political waters. A new era of mass politics seemed to be emerging out of the repressive shadows of the military republic. The inept political and economic maneuvers of the Figueiredo administration put the finishing touches on the now completely discredited military regime. Inflation had reached triple digits (much higher than in 1964), the foreign debt had passed $100 billion, and any claims that the country faced a leftist revolution or imminent political chaos had long since been discredited. In addition, as the regime had reached into every major sector of the economy, military officers had become involved in the operations of hundreds of corporations and agencies. Stories of military "perks" and military officers' ties to graft and corruption became commonplace in the increasingly aggressive and uncensored press. The discredited generals prepared to move back to the barracks, however reluctantly, and a civilian administration composed of the opposition and a faction of the ruling party prepared to take power. Brazil seemed poised to begin a new era.

A "NEW" REPUBLIC?

Unfortunately, the "New Republic" turned out to have many of the old patterns of Brazilian politics. Although the military stepped back from direct power, the generals remained highly visible and involved in the new regime. The deals cut by Tancredo Neves and his allies harkened back to long-standing traditions of the "old" politics of clientelism and patronage. In spite of the strong popular support for Neves and his movement, he came to power through a brokered and indirect election, and not through free and open elections. Finally, most of the major faces in the new government were part of the old regime. Neves, Sarney, and many others had started their political careers before 1964, and had remained important players in the political game throughout the military republic. In retrospect, the heady optimism on the eve of Tancredo's inauguration, the hope that a new day had dawned in Brazilian politics, seems especially misguided. The nature of the regime may have changed, but the patterns of politics had not.

The tragic death of Tancredo Neves on the eve of his inauguration disrupted the carefully brokered political deals. Without a Moses to lead Brazilians to the promised land, they now found themselves led by José Sarney, the Judas of the military regime, a man who had sold out his party in exchange for a vice-presidential nomination. Sarney's administration became a frenzy of patronage politics and corporate groups fighting to protect their interests.

The nation drifted while legislators drafted a new constitution and clashed over the scheduling of open presidential elections. In 1986, Brazilians went to the polls to elect a new congress that would simultaneously serve as a constituent assembly. The PMDB scored big gains in the elections just as it had in the races for state governors in 1984. While the PMDB controlled 22 of the 23 statehouses, it had 130 representatives out of 443 in the chamber of deputies. The Liberal Front Party (PFL), which had combined with PMDB to elect Tancredo Neves, controlled another 90 seats. Together, the two parties and a few other allies had a majority coalition. Ulysses Guimarães of the PMDB became the president of the congress/assembly. In his seventies, Guimarães was another politician who had survived the MDB's years as the official opposition, and whose political career extended back to the 1940s. He represented the continuity of the old politics with the new.

The drafting of the new constitution took two years and involved every significant interest group in Brazil. Everyone, from the generals to the taxi drivers, entered into the process to protect their group's interests. On the positive side, this produced debate and discussion that had been unthinkable in the authoritarian regime just a few years earlier. Under the scrutiny of a highly developed national media, the country could follow the emerging debates on the evening news or in the newspapers.

Yet beneath the surface of this "open debate," the enduring traits of Brazilian politics continued to operate. A clear hierarchy emerged as the traditionally powerful interest groups quickly established their preeminence in the assembly. Landowners, bankers, industrialists, and generals, for example, exerted enormous pressure on the legislators. Corporate interest groups, from the military to labor unions, fought to make sure their traditional rights were a part of the new charter. And, as always, the patronage system continued to operate with the state as the most powerful patron of all.

Although weakened, the central state still had more resources and favors at its disposal than any interest group in the system. A defector from the PDS, and distrusted by the PMDB, Sarney turned to the spoils of the state to influence legislators and to garner votes. He dispensed everything from government jobs to broadcasting licenses to buy votes. The struggle over the new constitution came to a head in mid-1988 over the issue of presidential term length (which would apply to Sarney and future presidents). Sarney fought bitterly for a five-year term and won after dispensing numerous political favors. The constitution was promulgated in October 1988, and direct elections for president were scheduled for November 1989. Unfortunately for Brazil, Sarney spent most of his term in office guaranteeing his own longevity and failing to give the nation any clear direction. For five years, the nation drifted.

Although the assembly contained delegates from virtually all sectors of the ideological spectrum, pragmatism prevailed. In the end, communists and conservatives, militant labor leaders and bankers, all brokered deals with each other to produce the new constitution. Almost as soon as the constitution went into effect, the campaign for president began, as the country moved toward the final step in the long process of *abertura*. Only with the inauguration of a freely elected president would the process be complete.

The spectacular rise of Fernando Collor demonstrated both the persistence of old political patterns and new forms of elite manipulation. Despite his campaign as an outsider and a new face, Collor came from an old land-owning family in the Northeast with a long tradition of political and economic power. (His maternal grandfather, Lindolfo Collor, was a key collaborator with Getúlio Vargas in the 1930s.) Young, handsome, and stridently critical of President Sarney and government corruption, Collor was a telegenic candidate. (The Collor candidacy has strong parallels to John F. Kennedy's presidential campaign in 1960.) Backed by Roberto Marinho, the owner of Brazil's most powerful communications conglomerate (*Rede Globo*), Collor quickly became the first Brazilian presidential candidate to be created by television. Fernando Collor's success made obvious to all the power of television in shaping national politics. With the major exception of Leonel Brizola, the older politicians did not fare well on the new political medium.

Collor emerged victorious with 53 percent of the vote. While Lula won in several of the cities, Collor rolled up an impressive vote in small cities and in São Paulo, Brazil's most populous state. He had very strong support from the middle and upper classes, but the poor gave him the margin of victory. Apparently Collor's style and telegenic presence enabled him to defeat Lula in the struggle for the vote of the masses. The biggest winner was the nation, when, as the third-largest democracy in the world, it staged a peaceful and fair election.

DRIFTING

Collor's fall was as swift as his rise (as seen in chapter 1), and it revealed the extreme corruption that permeates and pervades contemporary Brazilian politics. The traditional system of political patronage has always fostered a degree of graft and corruption. It has been built into the system. (Recently, a German watchdog organization, Transparency International, issued a ranking of corruption in forty-one countries, calculated from surveys of businessmen and journalists. Brazil ranked as the fifth most corrupt in the survey, behind Indonesia, China, Pakistan, and Venezuela.) Collor ran against the system, denouncing traditional patronage. In a perverse sense, he created a new patronage system. Instead of using the traditional lines of power and the state to dispense patronage, he made everyone come directly to him and his associates for favors. He completely personalized the patronage system at the very highest level of the political process.

P.C. Farias, his campaign treasurer, became the central figure in a huge kickback operation. Contractors, businessmen, and politicians had to arrange deals directly with Farias to gain concessions and assistance with their projects and programs. In exchange, they had to funnel money into a network of bank accounts that Collor and his cronies used for their own benefit. Collor's wife, Rosane, reportedly spent as much as $20,000 a month on jewelry and clothes, and the first family spent millions landscaping and renovating their home in Brasília. By some estimates, Collor and his associates siphoned off as much as $50 million through payoffs and kickbacks.

Fernando Collor's eventual impeachment in late 1992 represented both a victory and a condemnation of the system. On the bright side, the constitutional process worked. For the first time in this century, a major political crisis was resolved through constitutional means without the intervention of the military. Collor became the first democratically elected president ever to be impeached in Latin America—both a dubious and an impressive distinction.

A darker view of the process is that Collor became a sacrificial lamb, an offering to appease millions of Brazilians who were angered by political corruption on an unprecedented scale. Brazilian politicians embarked upon the impeachment process reluctantly, for they understood that intense scrutiny of the inner workings of the political process would reveal widespread corruption. Many powerful politicians wanted to use the impeachment process to pressure Collor to return to the old patronage politics—to set aside his maverick patronage system. As the process unfolded, these politicians got more than they bargained for, and they were forced by the logic of the process to remove the president from office. What began as pressure on the renegade president became a national political crisis. In the words of one politician, "We made a fire in order to roast a pig, but we end[ed] up setting the whole house on fire."

Over the past few years, a series of major scandals have rocked the political establishment. Investigations have revealed widespread payoffs and kickbacks to high-ranking politicians and government officials from nearly all parties. The great hopes of the eighties for a new democracy have soured into an enormous cynicism about politics-as-usual. Scandals have indicted the political process in the eyes of millions of Brazilians. The greatest danger to the "new republic" is not revolutionaries or generals, but a lack of faith in the system by the Brazilian voter. Honest and hardworking politicians have a tough task ahead convincing the voters that the system can be reformed.

In the last decade of the century, Brazil stands at a political crossroads. Having reestablished democratic forms of government, the nation's political leadership must now devise solutions to long-standing social and economic problems. The problems have grown to such proportions that they cannot be ignored or avoided. Brazil continues to face disadvantages as a late industrializer in a world dominated by powers in the Northern Hemisphere. As in the past, Brazilians face powerful disadvantages created by the structures of international politics and the world economy. Yet Brazil has become a middle power with a massive industrial infrastructure and representative government. Foreign influences do continue to shape domestic affairs, but the responsibility for Brazil's destiny is primarily in the hands of Brazilians.

Will the political leadership of the nation have the determination and the vision to begin to address the pressing social problems the country faces? Once again, the situation in Brazil parallels that in the United States. In both countries, deficit spending and foreign debt have made it difficult to marshal the resources to address fundamental social ills. In both nations, a pervasive cynicism makes voters wary about the system itself and its responsiveness to the "will of the people." The United States has been able to avoid the very tough choices that have faced the Brazilians, largely because foreigners (principally the Japanese and the Germans) continue to invest in the United States and to buy up debt. In the 1980s, foreign investors largely shut down the flow of new capital to Brazil.

Yet, more than a decade after the onset of the economic crisis and the end of military rule, the political leadership of Brazil has shown little desire to make fundamental changes. The old patronage networks and clientelism persist in new forms. Despite the return of open politics, the masses still do not have a powerful voice in a system that continues to be dominated by contending political elites. These elites still seem reluctant to address the fundamental changes so badly needed in a society with such profound social inequities—changes that, in the long run, would benefit both the elite and the masses.

Despite his glaring flaws, Fernando Collor initiated some impressive efforts to change Brazil. His moves to eliminate trade barriers and remove the government from many sectors of the economy have begun to move Brazil away from very deeply ingrained habits. Should the trend toward privatizing state corporations and eliminating most tariffs continue (a very big if), Brazil will have at least begun to reverse the centuries-old pattern of increasing economic centralization. These striking reversals, however, have been stalled by entrenched corporate interest groups, from labor to domestic producers who see their traditional interests threatened.

Collor also managed to disarm (both literally and figuratively) the military for the moment. Elected with the backing of very powerful conservative interests, Collor had the ability to say no to the generals on some of their demands, and to reassert civilian control over some traditional military prerogatives. In late 1990, he halted the secret nuclear weapons program of the military, publicly renouncing any ambitions to construct nuclear weapons. With a public relations flourish, he took the press to a secret site deep in the Amazon that the military had constructed for testing nuclear devices. Collor also appointed a renowned scientist, José Goldemberg, known for his criticism of Brazil's nuclear program, as his Secretary of Science and Technology.

Collor also made dramatic and sweeping pronouncements aimed at reversing the destruction of the Amazon rain forest and its remaining indigenous peoples. He shook up the Indian protection service, returned millions of acres of lands to Indian groups, and announced measures to slow the destruction of the rain forest. He appointed an internationally known environmentalist, José Lutzenberger, as his Secretary for the Environment. These efforts have been slowed or halted since 1992, but they clearly run counter to more than thirty years of government efforts to dispossess Indians of their lands, and to open up Amazonia to widespread exploitation. The fate of Indians and the rain forest remains up in the air.

Perhaps the most powerful challenge to any president is the deeply entrenched bureaucracy of the patrimonial state, what in Brazil is called "public functionalism" (*funcionalismo público*). The bureaucracy—federal, state, and local—employs about 6 million people, one out of every six salaried employees in Brazil. The salaries alone of public functionaries accounted for more than 60 percent of the federal budget in 1989. Historically, putting people on the public payroll has been a major means for politicians to attract and maintain the support of constituents. Legislation makes it very difficult to fire public functionaries, and this makes any effort to cut the bureaucracy problematic at best. The bureaucracy has long been the skeleton of the patrimonial state, and to attack it is to strike at the very nature of the system itself. Diminishing the power of the bureaucracy, and establishing account-ability in the system, are perhaps the greatest challenges facing any reformer in the long run.

Brazilian politicians could take lessons from Pope John Paul II in the revamping of a large, complex, and well-entrenched hierarchy. Over the past decade, the pope has initiated a very profound transformation of the institution that played such a prominent role in opposing the military repression, and in promoting redemocratization. The pope has been openly critical of liberation theology, and he has systematically silenced or replaced Brazilian church officials who aligned themselves with the activists. When visiting Brazil (for the second time) in October 1991, the pope made clear his commitment to social justice and called for fundamental social and economic changes in Brazil, but not through the methods of liberation theology. In the words of his chief spokesman, the Pope would like to see priests "illuminate the moral principles, not become political leaders." In his most visible move, he divided up Cardinal Arns' diocese in São Paulo, leaving him in charge of a much smaller diocese that includes some of the most affluent sections of the city. Within a period of a few more years, the Brazilian Catholic Church will have moved from being one of the most politically active and socially innovative in the Catholic world to a very conservative and quiescent institution.

CONTINUITY AND CHANGE

Despite the disastrous 1980s and early 1990s, the natural optimism of Brazilians (and this Brazilianist) is resurfacing. For although many of the old patterns and problems of Brazilian politics remain, Brazil has changed profoundly. Despite the elitism and hierarchy in Brazilian politics, the country is more democratic than at any other time in its history and certainly more democratic than most countries in the world today. The authoritarian and centralist regimes of the past have given way to more open and fluid politics. The state remains very powerful, and clientelism continues, yet a strong and aggressive press, and efforts to dismantle some state power, show signs that the nation may be able to move away from the centralist tradition.

The election of Fernando Henrique Cardoso to the presidency in late 1994 may signal the beginning of a new era in Brazilian history. On October 3, 1994, Cardoso easily defeated Lula, receiving 54 percent of the valid votes. With some 34 million votes, more than double Lula's tally, Cardoso has the largest electoral mandate in modern Brazilian history. As the foreign minister, and later finance minister, under Itamar Franco, Cardoso rode to victory on the success of the Real Plan, the economic plan that introduced a new currency in July 1994 and brought inflation down from 40 percent to 1 percent a month. He won the vote in every state except Rio Grande do Sul and the Federal District, and his party (PSDB) won the governorships of the three most powerful states (São Paulo, Minas Gerais, Rio de Janeiro). His coalition (with the PMDB and PFL) gives Cardoso a majority in congress and control of the governorships of fifteen states.

A sociologist with an international reputation, Cardoso has moved from the left of the political spectrum to a more moderate social democratic stance in the last twenty-five years. An intellectual with plenty of political experience, Cardoso has the skills to address Brazil's most pressing problems. His inaugural address on January 1, 1995, spoke to the need to confront the "social question," and during his first two years in office he has begun to push basic constitutional revisions through congress to revamp the economy and forge a national political consensus. With inflation at record postwar lows, a growing economy, and an emerging majority in congress, Fernando Henrique has the potential to move Brazil back onto the path of greatness. (With the passage of a constitutional amendment in early 1997 that allows the president to run for reelection, Cardoso also has the unprecedented opportunity to serve consecutive terms in office.)

The last years of the twentieth century will be decisive for Brazil. The nation has entered a period of enormous political flexibility. More so than at any other time in their history, millions of Brazilians can openly and freely make political choices that will determine their future. The Brazil of the 1990s looks very little like the centralist monarchy of the nineteenth century, in which a small elite disenfranchised the vast majority of the nation. Traces of the old politics remain, yet the long process of democratization during the past two decades has opened the way for all Brazilians to have a voice in the political arena. Brazilians must now find their voice, and speak clearly and authoritatively for fundamental change.

For two centuries, the Brazilians have moved cautiously and gradually toward more open and participatory politics. The emergence of democratic politics has not been inevitable or irreversible, nor has it been lacking in major setbacks. The process has been moved forward by a few enlightened leaders and enormous pressures from below, and by a pragmatism that has become a national trait. The elite's fear of open confrontation and conflict has generally moved them in the direction of conciliation and compromise, but they have not been able to avoid the emergence of stark inequalities that periodically threaten the future of all Brazilians. The political elite have rarely been polarized, but they have also failed to build a consensual center that would provide the country with stability and gradual reform. My own hope is that Brazil will be well served in the coming years by visionary, and pragmatic, leaders who will begin to address the basic social problems of the nation.

Ultimately, the "social question" must be addressed and answered in the political arena if Brazil is to become a truly modern nation. The challenges are formidable. The Brazilian masses must be enfranchised, both politically and socially, by reducing income inequalities and making sure that all Brazilians enjoy and exercise the full rights of citizenship. If Brazilians can muster the political will and leadership to address their pressing social problems, they will put the country back on the path to development, and Brazil could move into the ranks of the world powers in the next century as a strong, prosperous, and democratic nation. Should they fail to find a political solution, the country could descend into social chaos, condemning Brazil to remain the always-elusive country of the future.

FIVE

A Flawed Industrial Revolution

> Some people think that all the differences be-
> tween North Americans and us are economic. In
> other words, they are rich and we are poor; they
> were born in democracy, capitalism, and the
> industrial revolution and we were born in the
> Counter-Reformation, monopoly, and feudal-
> ism. But no matter how deep and determining
> the influence of the production system may be
> on the creation of a culture, I refuse to believe
> that our possession of heavy industry and our
> freedom from economic imperialism would suf-
> fice to erase our differences.
>   —Octavio Paz

BRAZIL BELIES THE WIDESPREAD BELIEF that economic growth and industrialization inevitably result in a higher standard of living for increasingly large numbers of people. Between 1870 and 1970, only Japan experienced higher economic growth rates than Brazil, yet today the majority of Brazilians live in poverty. From the 1930s to the 1980s, Brazil experienced something of an industrial revolution, becoming the tenth largest economy in the world and the most industrialized nation in the developing world. Yet Brazil now has the most inequitable distribution of wealth of any country in the world, and in the 1970s and 1980s the gap between rich and poor widened. Perhaps nowhere else in the world is the contrast between affluence and grinding poverty so striking and so ubiquitous. One Brazilian economist coyly suggested that the country be renamed "Belindia," because it has the advanced industrial base of Belgium with the social structure of India.

The flawed process of industrialization in Brazil contrasts sharply with the path followed by the industrial democracies of the North Atlantic community. England, Western Europe, and the United States industrialized in the eighteenth and nineteenth centuries in a world that was overwhelming agrarian. Brazil industrialized in this century and faced not only the competitive challenges of highly developed industrial economies in Europe, the United States, and Japan, but also the challenges presented by other newly industrializing nations. As social scientists have often pointed out, the later the onset of industrialization, the higher the costs of entry into the club.

In the older industrial powers, an enormous population shift accompanied the economic transformation. As industry in the cities grew, agriculture became more productive and efficient, and the rural population migrated into the cities. The increasing productivity of agriculture freed farm laborers to work in the factories. The attractions of city life, often accompanied by the forced removal of excess workers from the land, generated a process of rapid urbanization. Although these two processes (urbanization and industrialization) were neither gentle nor well coordinated, they moved forward roughly in conjunction with each other. In Brazil, as in most of the developing world, urbanization has outrun industrialization at a rapid pace. Industry has not been able to absorb the flow of migrants into the cities, and agricultural productivity has not kept up with the demand for food in the cities. The results are a still-backward countryside and cities surrounded by teeming slums of the impoverished and the underemployed.

The so-called rise of the middle class also accompanied urbanization and industrialization in Europe and the United States. Despite pronounced economic and social inequities, the size of the middle class gradually increased as cities grew and industry advanced. Although the middle class has grown in both absolute and relative terms in Brazil, the benefits of economic growth have been slower to "trickle down" to the masses than was the case in the North Atlantic world. Despite more than half a century of rapid industrialization, only about 15-20 percent of all Brazilians can be classified as middle class (as opposed to about 60 percent of all North Americans). Ironically, this small middle class has become a major stumbling block in continuing the process of economic and industrial growth in late-twentieth-century Brazil.

Clearly, Brazil presents a powerful challenge to those who would argue that economic growth built on industrialization in the developing world will reproduce the widespread affluence and democratic politics so characteristic of Western Europe, the United States, and Japan. The equating of industrial revolution with middle-class society and democracy does not fit the Brazilian experience. It is a mistake to assume that capitalism will develop the same way, and produce similar results, at all times and in all places. Capitalism has been firmly entrenched in Brazil for decades, if not centuries, yet this "tropical capitalism" has evolved in ways never experienced above the Tropic of Cancer.

THE COLONIAL HERITAGE

A good deal of the explanation for this flawed process of economic growth lies in Brazil's colonial heritage, although the burden of that heritage has been eased more dramatically in the economy than in contemporary politics or society. The patterns developed during three centuries of colonial exploitation have been altered very slowly, but they have been altered. The dismantling of the colonial economic heritage began in the mid-nineteenth century and accelerated at the turn of the century. Since the 1930s, a flawed industrial revolution has profoundly transformed the Brazilian economy.

The colonial economic heritage can be summed up in one sentence: The colonial economy was dominated by large, landed estates, and controlled by a small white elite, who used black and Indian slaves to produce export crops or precious metals for European markets. This should not be surprising. After all, the purpose of a colony is to supply the mother country with riches. By its very nature, a "successful" colonial economy is extractive and export-oriented, producing wealth for the metropolis. While New England failed, in this sense, as a colony and turned to commerce and shipping as its role in the British colonial system, Brazil "succeeded" in spectacular fashion. Consequently, New England failed to develop the most pernicious traits of the "successful" colony, the traits that defined Brazil.

The Brazilian colonial economy also moved through a series of cycles that all relied on the overwhelming dominance of a single product: first brazilwood, then sugar, then gold. Each cycle roughly corresponded to a century. The Portuguese cut and shipped the dye-producing brazilwood throughout the sixteenth century. Sugar production exploded at the end of the sixteenth century and went into decline roughly a century later. Prospectors discovered gold in the 1690s, and the next sixty to seventy years were Brazil's "golden age." With the rapid decline in gold production after 1770, Brazil continued to ship brazilwood, sugar, and gold to Europe, but no single product dominated. The next cycle would begin with coffee, but not until the 1830s.

COFFEE AS THE CATALYST

The coffee cycle would run for more than a century, but unlike brazilwood, sugar, and gold, this export became an engine of economic growth that contributed to industrialization. Coffee became an agent of change that contributed to the decline of slavery and the rise of free labor. The expansion of coffee cultivation attracted millions of immigrants to the Southeast and transformed the city of São Paulo from a small, frontier town into the largest industrial center in the developing world. The rise of the coffee economy, first in the province of Rio de Janeiro, and then São Paulo and Minas Gerais, continued the shift in power begun with the discovery of gold in the 1690s. Gold made the Southeast the locus of economic and political power in Brazil. Coffee consolidated the rise of the Southeast and the decline of the Northeast.

For those who would argue that monocultural export economies are locked into economic dependency on the manufacturing nations of the North, Brazil once again belies theory. Despite Brazil's overwhelming reliance on coffee exports (which accounted for as high as 90 percent of the value of all exports in some years in the nineteenth century), rather than falling into an inescapable form of economic dependency, the coffee economy generated what economists like to call "spread effects" and "linkages." Entrepreneurs reinvested profits from coffee into banking, commerce, and industry, thus stimulating the growth of various sectors of the economy. Many of the early industrial entrepreneurs came from landholding families, and the immigrants attracted to the Southeast played a fundamental and catalytic role in the origins of industrialization. Coffee helped expand the use of free wage labor and the growth of a domestic market. Finally, it is not coincidental that the most dynamic industrial center in the developing world arose in the heart of the coffee-growing Southeast.

This is not to say that coffee acted only as an agent in the modernization of the Brazilian economy. The process was complex. For while the coffee economy spurred modernization, in many ways it also reinforced the traditional features of the colonial heritage. The coffee barons may have been a new economic elite, but they were still an elite, and a landholding one at that. By no stretch of the imagination did all coffee planters promote free labor, immigration, and the abolition of slavery. Even the most entrepreneurial and "progressive" planters took a cautious approach to tinkering with the basic structures of a society constructed on slave labor and profound class differences. The Brazilian elite debated the advantages and disadvantages of slavery and free labor for decades and took the most conservative approach to legislating abolition of any political leaders in the hemisphere. Only after enormous pressure from the British, the threat of chaos at home, and the beginnings of a massive wave of European immigration did the Brazilian parliament finally abolish slavery in 1888.

ELITES AND INDUSTRIALIZATION

As Brazilian politics clearly show, the ideology and values of the Brazilian elite have repeatedly blocked efforts to turn to new forms of political, economic, or social organization that have been proven and tested elsewhere. To be more specific, the Brazilian elite had the option to industrialize (recognizing that the shift would not be easy nor unproblematic), but they consciously chose to follow an agrarian model. They debated about what kind of economic policies best suited the nation throughout the first century of independence. The view that prevailed during most of that period was that the country was "essentially agricultural" and that economic policies should build on that strength. Captivated by classical liberal economics, the predominantly landholding and agrarian elite pursued what twentieth-century economists would call their "comparative advantage."

For well into this century, a minority (elite and non-elite) called for industrialization. Only near the end of the nineteenth century did they begin to make headway, and not until the shattering experience of the Great Depression did their viewpoint begin to prevail. This minority fought an uphill battle in an economy and polity overwhelmingly dominated by rural, landholding elites both in the Northeast and the Southeast, the two power centers in the nation. Even after the Great Depression exposed the fragility of an economy built on the export of a non-essential crop, the proponents of industrialization could not gain the upper hand. As late as 1960, coffee exports still accounted for 60 percent of the value of all Brazil's export. Despite several decades of impressive industrialization and government support for industrialists, the Brazilians remained heavily dependent on a single export crop.

The debates over economic policies illustrate the ideological contradictions that also plagued Brazilian politics. The centuries-old Iberian corporatist ethos that continued to shape the elite worldview made it difficult to wean them from protectionism and interventionism. At the same time that many of the founding fathers embraced classical political liberalism, many also found economic liberalism appealing. This fervent attachment to economic and political ideas forged in the North Atlantic world forced the Brazilian elites to perform intellectual gymnastics. They supported individual liberties and equality before the law, but only for those who met the requirements of citizenship—that is, propertied, literate, adult, "white" males, like themselves. They believed in free trade and laissez-faire, yet government protection of traditional monopolies and intervention into key economic sectors proved hard to abandon.

Ironically, the rhetoric of economic liberalism eventually triumphed in nineteenth-century Brazil, while government intervention, monopolies, and protectionism quietly continued, and grew. As one prominent economic historian has pointed out, during the First Republic (1889-1930), a regime that proudly espoused its liberalism, the Brazilian elite constructed one of the most interventionist regimes in the Western world. The mark of genius, someone once commented, is the ability simultaneously to hold two mutually contradictory ideas in one's head. Brazilian elites have demonstrated true genius for at least a century, passionately espousing liberal economics while refusing to part with their attachment to state intervention.

Until the twentieth century, the state normally intervened to protect traditional prerogatives and economic activities, not to foster the creation of new industry. In the nineteenth century, the interests of landowners and those tied to the export agriculture economy took precedence over the interests of the tiny industrial sector. Foreign manufactured goods, principally British manufactures, dominated the Brazilian market. The British could mass-produce cheap, quality goods, ship them around the globe, and still undersell local producers. The enormous advantages from economies of scale and the political backing of their government helped the British sweep away local competition.

EARLY INDUSTRIALIZATION

In spite of the tremendous competitive pressures from foreigners, domestic producers survived and grew, albeit slowly and in limited sectors of the economy. Throughout the nineteenth century, Brazilian manufacturing grew slowly, producing goods for consumption in local and regional markets: textiles and clothing, iron, furniture, and processed foodstuffs. In these areas, principally what economists call nondurable consumer goods, local producers retained something of an advantage over foreigners. These goods could be produced with relatively simple technology, in cottage industry, and to satisfy markets (especially in the interior) that were not easily accessible from major ports. By the end of the nineteenth century, Brazil had a small, but growing, industrial base.

Most of this production developed in the more populated areas of the Southeast, the Northeast, and the Far South. Salvador, Recife, and Porto Alegre had small industrial establishments. In the interior of Minas Gerais, dozens of small manufacturers produced goods for consumption in regional markets. The industrial growth in the cities of Rio de Janeiro and São Paulo, however, easily overshadowed the industrial growth in all other regions of the country. As the national capital and principal port, Rio de Janeiro became the logical location for early industry. By 1890, the city had a population of more than half a million inhabitants, a considerable consumer market.

The spectacular growth of the city of São Paulo, and its emergence as the largest industrial center in the Third World, could not have been predicted by anyone in the nineteenth century. A rustic town located just over the high coastal escarpment, São Paulo had fewer than 30,000 inhabitants in the 1850s. With the expansion of coffee cultivation into the surrounding hinterland in the late nineteenth century, the city became a dynamic and booming *entrepôt*, connected to the port of Santos by a railway. Between 1890 and 1900, the city's population grew from about 70,000 to nearly 240,000. Tens of thousands of immigrants—Brazilians, Italians, Germans, and many others—flooded into the city and the state. The entrepreneurial talents of coffee planters, Brazilian industrialists, and immigrants transformed the city, creating a magnetic pole of industrial growth. By the 1930s, the state had become the most populous in Brazil, the city had a population of over a million, and the city of São Paulo moved past Rio de Janeiro as the most important industrial center in the nation.

Brazil's industrialists benefitted somewhat from the disruptions that World War I produced in the Western economy. The war disrupted the normal trade flows from the North Atlantic, making it difficult for manufacturers to import essential technology. The scarcity of imports helped local producers of consumer goods, who expanded output of textiles, clothing, shoes, and processed foods. The war also demonstrated to many of the elite the nation's economic vulnerability. They could see how dependent their economy was on the European coffee markets, and how much they depended on European and North American imports. Industry emerged from the war strengthened, and industrialists found a larger and more receptive audience when they voiced their concerns.

In the 1920s, the Brazilian economy continued to expand and diversify, but its dynamism still hinged on coffee exports. In the mid-twenties, coffee accounted for three-quarters of the value of all exports, and nearly 10 percent of the Gross Domestic Product. The coffee states of São Paulo, Rio de Janeiro, and Minas Gerais continued to dominate national politics, much to the dismay of those, such as the *tenentes,* who called for a rapid shift toward industrialization. While the political elite were not simply the representatives of coffee interests, they did recognize the centrality of coffee in the national economy.

The political elite carefully protected the coffee economy, instituting a "valorization" scheme in 1906. Confronting a glutted world market, the government set up a price-support system to buy excess production. The government hoped that this interventionist measure would be temporary, keeping producers in business during low-price periods and driving up prices by keeping supplies off the market. The coffee valorization plan forced the government into deficit spending and external borrowing, and it ultimately failed. Overproduction was not a temporary phenomenon, and many of the stocks had to be destroyed rather than eventually placed on the market.

VARGAS AND INDUSTRIALIZATION

The Great Depression sent powerful shock waves through the Brazilian economy and forced the elite to rethink and reorient economic policies. Between 1929 and 1932, the value of exports dropped by 60 percent as world trade nosedived. The influx of foreign capital halted and essentially forced Brazil to suspend payments in 1931 on more than $1.3 billion in external debt. Much as it would in the 1980s, Brazil began a long, drawn-out process of renegotiation with foreign lenders to restructure and consolidate its foreign debt.

As in the United States, the shattering economic effects of the Depression forced the government to take interventionist measures that had been unthinkable just a few years earlier. As the preceding account emphasizes, major government intervention in the economy did not begin in the 1930s, but the decade clearly marked a momentous shift in the trajectory of the Brazilian economy. Although the first impulse of political leaders was to prop up and protect the coffee economy, in the 1930s a shift began toward increasingly supportive moves to stimulate and develop domestic industry. The movement toward economic diversification built on industrialization using government intervention began in these years, eventually climaxing in the massive industrialization and intervention of the 1970s.

Forced by circumstance to reexamine the nature of the economy, Brazilians continued the debate over laissez faire and interventionism with a decidedly new tilt in the direction of the latter. In the 1930s, for the first time, the government made important efforts at economic planning and at systematically assessing the state of the economy. Policymakers resorted constantly to tinkering with all the fiscal and monetary measures at their disposal, thus creating new agencies and mechanisms for intervention. Just as in the United States and Europe, "priming the pump" with Keynesian economic policies became standard procedure. (In a sense, the Brazilians had become Keynesians before Keynes with the coffee valorization scheme in 1906.)

As the head of state from 1930 to 1945, Getúlio Vargas played a central role in the new interventionism. The Vargas of these years, however, was a reluctant interventionist who kept one foot in the old liberal school of economic theory. Although many of the measures taken by the Vargas governments during these years stimulated industrial production, much of the resulting industrial expansion came as side effects of policies that did not directly aim to stimulate domestic industry. This has been called "spontaneous import-substitution industrialization," as compared with the more managed and directed efforts to substitute imports with domestic production in the post–World War II period.

By the end of the war, Brazil imported just 20 percent of its industrial products, with domestic industry producing the 80 percent of national needs. In the area of light consumer durables, such as furniture, household goods, and clothes, national industry supplied close to 100 percent of domestic demand. As goods produced by domestic industry gradually replaced foreign manufactures, Brazil's balance-of-payments picture improved. Fewer imports and growing exports produced a positive balance-of-payments and allowed the government to build up a substantial surplus of foreign exchange reserves.

This industrial expansion, however, was somewhat illusory. As the eminent economist Werner Baer has pointed out, most of the industrial growth during this period came from utilizing preexisting industrial capacity, not from the creation of new industrial infrastructure. The expansion during the 1930s and World War II did not produce new industry, but more fully utilized old industry. Consequently, when the war ended in 1945, and international trade relations normalized, much of Brazil's industry became outmoded and obsolescent.

By 1945, Brazil's economy had become more diversified and industrialized, but it remained heavily dependent on agricultural exports. The agricultural sector accounted for 28 percent of the Gross National Product (versus 20 percent for the industrial sector) and more than *60 percent* of the economically active population worked in agriculture. Agricultural exports (coffee, sugar, cocoa, cotton, tobacco) still accounted for the bulk of Brazil's exports. The terms of trade, as economists would say, declined in the postwar years as the prices of Brazil's agricultural exports declined relative to the cost of the imports of capital goods and fuel needed to power the industrial economy.

DEVELOPMENTAL NATIONALISM

Brazilian policymakers recognized that future economic growth hinged on increased dynamism in the industrial sector and decreased dependence on agricultural exports and industrial imports. They sought to change the nature of the economy by consciously promoting import-substitution industrialization. Since 1945, economic policymakers have pursued these goals, albeit with very diverse approaches and ideological agendas. The drive to industrialize has transcended divisions among political parties and regimes. Under a democratic regime in the 1950s, the Brazilian economy surged forward with impressive industrialization. Under a dictatorial regime in the 1970s, the nation experienced a spectacular industrial spurt that became known as the Brazilian "miracle." In the early 1950s, Brazil's economy was the fiftieth largest in the world. By the late 1970s, it had become the tenth largest.

In the 1950s, the move toward state intervention to promote industrialization crystallized. Known in Portuguese as *desenvolvimentismo* (developmentalism), it was a mentality shared by most of the Brazilian elite. They often disagreed violently on how to achieve development, but virtually all elite groups believed that Brazil's future depended on the promotion of industrialization. The debate over the extent of government intervention in the economy became an issue that bitterly divided Brazilians. In general, those on the political left favored more nationalistic, protectionist, and interventionist policies, while those on the right favored more traditional laissez faire approaches to industrial promotion. Throughout the 1950s and early 1960s, nationalism and interventionism prevailed.

Economic nationalism, state intervention, planning by technocrats, and foreign investment all converged to produce very high economic and industrial growth rates. From 1940 to 1980, real output increased every year (except 1942), and the Brazilian economy grew at a rate of 7 percent a year. Not only did the Gross Domestic Product

(GDP) grow, real GDP per capita quintupled in the same period to nearly $2,500, one of the highest rates of growth in the world during these four decades. During the presidential administration of Juscelino Kubitschek (1956-61), economic growth surged. This brilliant political improvisor juggled the interests and methods of both left and right to pursue his plan of moving Brazil forward "fifty years in five." Kubitschek placed technocrats and planners in positions of power in his administration. As soon as he took office, he created a National Development Council, and his technocrats issued a now-famous "Program of Targets" *(Programa de Metas)*. The program identified key sectors of the economy and set growth targets for both government and the private sector.

Under Kubitschek in the late fifties, overall industrial production rose by 80 percent, while in several key industries (steel, mechanical, electrical and communications, and transportation equipment), growth rates ranged from 100 to 600 percent. Brazil moved dramatically from being mainly a producer of non-durable consumer goods toward the goal of producing its own intermediate and capital goods. The creation of a powerful automotive industry was the most dramatic example of Kubitschek's industrial promotion. When Kubitschek entered office in 1956, Brazil manufactured very few automobiles. By 1960, the automotive industry (led by Volkswagen) produced more than 130,000 vehicles, with more than 90 percent of the manufacturing content coming from Brazil.

By 1960, manufacturing accounted for a greater share (28 percent) of the GDP than agriculture. Even when adjusted for an annual population growth rate of 3 percent, Brazil's per-capita real growth in the 1950s was about three times that of the rest of Latin America in a period when the region had the highest growth rates in the developing world. Brazilian industry in the early sixties supplied nearly 100 percent of the country's consumer goods, about 90 percent of its intermediate goods, and close to 90 percent of its capital goods. The industrialization of the 1950s, especially under Kubitschek, laid the groundwork that made possible the "miracle" of the late sixties and early seventies.

At the very moment that Brazil seemed poised for a major industrial "takeoff," the country entered into the severe political and social crisis of the early sixties. The impressive economic surge of the fifties slowed down and then halted. Kubitschek's expansionary program had been financed through deficit spending, and the government debt had been financed by printing more currency and by foreign borrowing, both inflation-generating measures. By 1960, inflation rates had passed 20 percent per annum, and the nation faced a severe balance-of-payments crisis. International lenders and lending agencies pressured Kubitschek to impose an economic austerity program to lower inflation, cut imports, expand exports, and cut government spending. Juscelino refused to resort to austerity. He wanted to be remembered as the president who built the new Brazil, not the president who imposed economic hardship.

The triple threat of inflation, deficit spending, and a balance-of-payments crisis threatened to unravel the impressive economic gains of the 1950s. Chronic double-digit inflation created social discontent, particularly among the politically volatile middle class. Deficit spending financed growth but fueled inflation and forced the government to borrow abroad to finance continued economic expansion. External borrowing, ultimately, hinged on foreign lenders' confidence that the Brazilian economy would continue to grow and offer a stable investment climate.

Even the politically adept Juscelino finally ran into trouble facing this three-fold threat. When he left office in January 1961, he handed these complex problems over to Jânio Quadros and João Goulart, who turned them into a monumental political and economic disaster. By early 1964, inflation approached 100 percent on an annual basis, foreign investment had halted, and Brazil had effectively defaulted on its $2.5 billion external debt. Economic growth stalled. Goulart's mismanagement of the economy provided one of the major motivations for military intervention in 1964.

TECHNOCRATS AND GENERALS

Just as the military had developed a vision of how they wanted to revamp the nation's political system, they had also long sought to transform the nature of the Brazilian economy. For nearly a century, military officers had been vocal supporters of the drive to industrialize. They recognized that the armed forces could be (and often were) held hostage to the whims of the industrial powers that supplied them with equipment and arms. The officer corps understood that a strong domestic industrial base would reduce their dependence and strengthen the armed forces. Military officers also clearly recognized that no nation could achieve power in the world community without a strong economy, and in the twentieth century that translates into a strong industrial base. For decades, the armed forces promoted industrialization as both a means to strengthen their own power and to make Brazil more powerful in the international arena.

Once again, the legacy of positivism permeated the military's mentality. In the aftermath of the coup in 1964, the high command would turn to technocrats to restructure and run the economy as an authoritarian republic. The technocrats would not have to contend with the pressures of an open political system. While the generals insured political order, the technocrats pursued economic progress. Freed from many of the constraints that economic policymakers have to contend with in democratic polities, the technocrats set about the business of transforming the Brazilian economy.

Under the military, the Brazilian economy initially stagnated; then experienced an unprecedented boom, followed by continued high growth accompanied by growing problems; and then slipped into a prolonged crisis. As Brazil rode this economic roller coaster, the generals loudly proclaimed their faith in capitalism while continually extending the role of the state in the economy. The nationalistic right-wing generals created the most statist economy in the non-communist world. By the 1980s, the state accounted for more than 60 percent of Brazil's Gross Domestic Product.

Initially, the military regime turned to a fairly orthodox approach to economic policymaking. Roberto Campos and Octávio Gouvéia Bulhões, both respected economists with strong ties to the banking and business communities, headed the economic team of the new regime. In contrast to the left, which argued that capitalism had failed in Brazil, they believed that capitalism had never been given a chance to work. They wanted to open up the economy to foreign investment, reduce inflation, cut the government's deficit, and modernize the capital markets and financial institutions that are essential for entrepreneurs in a capitalist system. Rather than forcing industrialization through state intervention, they hoped to develop the export sector and generate growth that would not have to be state-directed.

Freed from the political pressures that had faced Kubitschek, Quadros, and Goulart, Planning Minister Campos imposed the kind of tough (and very unpopular) orthodox stabilization program his predecessors had been unwilling or unable to risk. Campos and Bulhões slowed down the printing presses and the expansion of the money supply. More rational and efficient tax collection helped cut the government's budget deficit. Adjusting the exchange rate promoted the expansion of exports. Along with declining demand for imports, this eased the balance-of-payments problems. By 1967, they had succeeded in reducing the government deficit, and inflation had dropped from 100 to 27 percent per annum (although they did not reach the announced target of 10 percent). They succeeded in reopening the flow of foreign investment, especially in the export sector. They also achieved some success in fostering the growth of capital markets in the form of investment banks, credit funds, and stronger stock markets.

Despite the drop in inflation, and the improving investment picture, the economy continued to stagnate between 1964 and 1967. Efforts to bring down the inflation rate had prolonged the recession that had begun under Goulart. From 1962 to 1967, economic growth rates averaged 3.4 percent per annum, easily the lowest rates of growth in the postwar decades. In spite of the low growth rates and recession, Campos and Bulhões had laid much of the groundwork for the phenomenal growth that would take place over the next seven years. They had brought inflation down dramatically and reduced the inflationary pressures created by deficit spending. By controlling wages and labor, they further reduced inflationary pressures. Perhaps most importantly, they had initiated a fundamental shift away from import-substitution, with its focus on developing the domestic market, and moved the nation toward the promotion of exports (agricultural and manufacturing) and the development of external markets for Brazilian goods.

As Costa e Silva and the hardliners took over from the Castello Branco administration in 1967, the right-wing nationalists were impatient with liberal economists and their gradualist approach. For the staunch nationalists, the liberal economic policies did not move Brazil quickly or vigorously enough into an industrial future. These impatient officers chose increasingly statist and interventionist policies to move the economy forward dramatically and quickly. Over the next decade, these policies, combined with extremely favorable circumstances in the international economy, would produce the most impressive period of economic growth in Brazilian history. While the generals unleashed the security apparatus to pursue internal political order with a vengeance, they let the technocrats zealously extend the reach of the state into the economy in search of economic progress.

Ultimately, the path pursued by the technocrats avoided the extremes of both the free-marketeers and the radical nationalists. Led by the young *paulista* economist, Antônio Delfim Neto, the regime embarked on a pragmatic path that welcomed foreign investment *and* sought to develop domestic industry. In effect, these pragmatists agreed with the nationalists on the need to develop manufacturing in Brazil, but unlike them, they were not concerned with who owned the plants. For the pragmatists, "made in Brazil" was not the same as "made by Brazilian-owned industry." Like the liberals, Delfim Neto welcomed foreign investment with open arms, yet the state played a crucial role in attracting, nurturing, and guiding foreign capital into the sectors of the economy that the government considered critical for national economic growth.

Pragmatic economic policies and favorable conditions in the world marketplace converged to produce the so-called economic miracle from 1967 to 1973. Rising demand for Brazil's exports and easy credit from international lenders created optimum conditions for economic expansion. World trade, partly stimulated by the war in Vietnam, expanded at nearly 20 percent per year in the late sixties and early seventies. Brazil's agricultural output and productivity rose under the most favorable international circumstances since the Korean War. The rapidly growing economies of Europe and Japan generated large amounts of funds for investment, and the military regime made Brazil an attractive place to invest. Multinationals alone invested more than $6 billion in Brazil during these years, a very large sum for a country that imported only about $3 billion worth of goods each year.

Delfim continued many of the policies of Campos and Bulhões. He also believed in the outward-oriented growth model promoting the expansion of both agricultural and manufacturing exports for foreign markets. He developed a policy of small and continual devaluations of the Brazilian currency (*cruzeiro*) to keep it at a realistic level with foreign currencies. This policy of "crawling peg" or mini-devaluations kept Brazilian products from becoming too expensive for foreign buyers. Like Campos, he tried to make Brazil as attractive for foreign investors as he could.

Delfim did diverge from his predecessors in two important ways. Much like the supply-side economists of the 1980s, he wanted to stimulate demand by making credit cheaper and more readily available. Campos had implemented an austerity program to dampen demand, believing that this was the only way to lower the inflation rate. Delfim loosened up credit and increased the money supply and the economy took off in 1968-69. Contrary to the predictions of Campos and many others, inflation decreased between 1967 and 1972, even as demand and production rose.

Campos and Bulhões had reluctantly resorted to measures such as price controls in the short-term. Delfim and his economic team adopted intervention as a long-term strategy. Price controls were aimed at dampening inflation. Wage controls helped reduce inflation and also forced workers to pay the price through a reduction in real income over the succeeding years. The government systematically kept wage hikes behind inflation through a process known as "monetary correction," or indexation. Begun in a small way under Campos, indexation soon extended into all sectors of the economy. In what would become a difficult-to- dismantle nightmare, the government tried to tinker with all sectors of the economy by establishing formulas for readjusting prices, wages, and bank accounts. Periodically, the government announced the official figures for "correcting" bank accounts, prices, and wages. In effect, the government itself generated inflation through the constant readjustment of indexes.

THE "MIRACLE"

From 1967 to 1973, the spectacular growth of Brazil truly seemed "miraculous" to many observers. In addition to the very high growth rates (averaging nearly 11 percent per year), dropping inflation rates (to around 17 percent per year), and continually rising domestic demand and output, the traditional problem with the balance of payments seemed to have been solved. In 1969, Brazilian banks had accumulated about $650 million of foreign exchange holdings. This amount grew by a factor of ten by 1973 to nearly $6.5 billion, a phenomenal increase. Export promotion dramatically increased the total value of exports from $2.7 billion in 1970 to $7 billion in 1973. The influx of foreign capital that had come to a virtual halt in 1964 surpassed $4 billion in 1973. No wonder that by mid-1973 Delfim Neto seemed to many to be an economic genius.

Delfim Neto and the technocrats pushed Brazilian industrialization forward dramatically and impressively, yet they were able to do so because of the decades of industrial growth that had preceded the military regime. Vargas and Kubitschek, in particular, developed three crucial industries that would become basic pillars of the industrial spurt after 1967: iron and steel, automobiles, and electric power generation. By the sixties, Brazil had become one of the world's leading producers of iron and steel, providing essential inputs for Brazilian industry, as well as valuable export earnings. Although in the 1940s Vargas had pushed the construction of South America's first integrated iron and steel complex at Volta Redonda (between Rio de Janeiro and São Paulo), the state of Minas Gerais had become the heartland of production.

With some of the largest iron ore reserves in the world, Minas Gerais was the logical place to build iron and steel plants. Small foundries, financed largely with local capital, had sprung up in Minas Gerais in the nineteenth century, but it would take large amounts of foreign and public financing after World War I to exploit the massive ore reserves. Local investors could not mobilize the capital, nor could they compete with the powerful producers in Europe and the United States. In the 1920s, European investors, anxious to gain access to the valuable ore reserves, formed the Belgo-Mineira Iron and Steel Company near Belo Horizonte. Like many smaller, locally owned companies, Belgo-Mineira fueled its furnaces not with expensive coal imported from overseas, but with wood from the surrounding forests. By World War II, Minas produced about three-quarters of Brazil's iron and steel.

The military accelerated the exploitation of the enormous iron ore reserves in Minas Gerais by bringing in more foreign capital and by investing large amounts of public monies. The Nippon Steel Corporation of Japan made large investments in an iron and steel complex in southern Minas Gerais at Ipatinga. A joint venture of primarily Japanese and *mineiro* state capital, Usiminas (Usinas Siderúrgicas de Minas

Gerais) became a very successful enterprise that was run efficiently and profitably by Brazilian technocrats and Japanese businessmen.

To the northeast of Belo Horizonte, the military regime chose to maintain national control of the huge ore reserves and production. Under the military, the state-controlled CVRD (Companhia Vale do Rio Doce, created in 1941) became the largest iron mining company in the world. The CVRD developed the region's small iron and steel industry into an enormous iron-mining, and iron-and-steel-producing, center. The so-called Valley of Steel became the hub of mining ore for export, and of forging iron and steel for both national consumption and export overseas. Brazilian steel production surpassed that of both France and England. In the early sixties, Brazil exported about 5 million tons of iron ore a year. By the 1980s, iron ore exports approached 100 million tons and accounted for 8 to 9 percent of all Brazil's exports. In the 1970s, the CVRD and Usiminas became outstanding examples of the use of state funds to promote vital industries that were both efficient and profitable. (*Fortune* magazine singled out the CVRD in 1989 as the most profitable corporation of the 500 largest corporations in the world. According to some estimates, the company made a $40 profit for every $100 of revenue.)

Steel production and exports rose rapidly under the military regime. By the mid-1980s, Brazil had become the largest exporter of steel products in the developing world, and it accounted for 4 percent of world exports. Exports of steel brought in more than $2 billion in foreign currency, thus making steel Brazil's most important manufacturing export. The United States and Latin American nations imported nearly two-thirds of all the steel products Brazil exported in the 1980s, although Japan has become an increasingly important market during the past decade. Although foreign capital played a crucial role in the growth of this basic industry, private Brazilian capital or public monies controlled the majority of the steel-producing companies.

A rapidly growing automotive industry consumed much of the iron and steel produced by Brazilian foundries. One of the great success stories of import-substitution industrialization policies, the Brazilian automotive industry grew from very limited production in the 1950s to be the eighth largest producer in the world by the 1970s. In the late fifties, Brazil produced just 30,000 vehicles per year, containing over 50 percent imported components, and none of these were passenger vehicles. By the mid-1980s, the Brazilian automotive industry manufactured 1 million vehicles annually, with close to 100 percent local content. The mix in production had also shifted. Passenger cars accounted for three-quarters all vehicle production by 1985.

Massive foreign investment, first under Kubitschek, and then under the military regime, built the automotive industry. Juscelino established special incentives for foreign manufacturers to build their plants in Brazil, and gradually to increase the percentage of parts manufactured in Brazil. This produced what economists call "vertical integration," that is, by the 1970s a variety of industries had taken shape in Brazil, each producing parts that converged to produce the finished product: a Brazilian-made automobile. Kubitschek courted both North American and European manufacturers assiduously, yet only Volkswagen had the foresight to make a big entry into this untapped market.

Investing hundreds of millions of deutsche marks in the construction of plants, the Germans quickly became the overwhelmingly dominant force in automotive markets. The Volkswagen "beetle" (or *fusca,* as it is called in Portuguese) became *the* Brazilian car. As late as the 1970s, one could travel for miles on Brazilian highways without passing anything but a Volkswagen. In some cities, *fuscas* made up the entire taxi pool on the streets. Volkswagen remains the dominant force in the industry, producing about half of all passenger cars manufactured in Brazil.

During the heady boom period of foreign investment in the 1970s, General Motors, Ford, Chrysler, and Fiat also became major manufacturers. These multinationals hoped to create "export platforms" in the developing world. By building plants in Brazil, for example, the multinationals avoided the high tariffs and duties on vehicles imported from the United States or Europe. The plants in Brazil would have an inside track on the domestic market in Brazil, and would serve as a form of "platform" for launching into the markets of other developing nations.

The automotive industry became a key sector in the national economy, employing about 100,000 workers in large factories that primarily concentrated in the industrial belt of Greater São Paulo. The ABC region (Santo André, São Bernardo, and São Caetano) to the south of the city of São Paulo became the center of Brazilian auto manufacturing and the home of the densest concentration of industrial workers in the country. In the late 1970s, Fiat constructed a major assembly plant on the outskirts of Belo Horizonte. The plant employs tens of thousands of workers and produces upwards of 100,000 vehicles annually.

Although more than 90 percent of the components used in manufacturing vehicles are Brazilian made, the automobiles are decidedly "foreign." European, North American, and later Asian capital financed the construction of the automotive industry, and these countries ultimately control it. The high costs of investing for these foreign manufacturers translated into high prices for the products. In the 1970s, the average price of a passenger car in Brazil ran nearly double the average price in the United States. The manufacturers recouped their investment in the form of high profits. (In the 1970s, the foreign units of North American automotive manufacturers were the most lucrative branches of these huge multinational corporations.) The Brazilian government recouped much of its enormous investment in the form of high taxes. (Taxes and duties account for about half the price of Brazilian passenger cars.)

In the 1970s and 1980s, U.S. and European firms took divergent approaches to developing their subsidiaries in Brazil. U.S. auto firms produced finished vehicles for the Brazilian market but did not attempt to export these vehicles back to home markets. Instead, the U.S. subsidiaries in Brazil became suppliers of auto parts and engines for North American factories. General Motors and Ford increasingly moved components manufacturing to foreign locations to take advantage of cheap labor, and then imported the parts from their subsidiaries to be assembled in U.S. plants. Volkswagen, on the other hand, pursued a strategy of increasing the export of finished vehicles to home markets. Volkswagen also began an effort to develop a U.S. market for Brazilian-made vehicles. The Volkswagen Fox and Passat, both developed and manufactured in Brazil, appeared in the U.S. auto market in the late 1980s.

In developing the Brazilian automotive industry, planners consciously fostered the creation of a transport system built around trucking and an enormous highway system. The construction of national highways (much like in the United States) first promoted by Kubitschek in the 1950s was aimed at integrating the nation and developing a petroleum-based transport system. In the days of cheap oil, this seemed like a reasonable plan. After 1973, this development model began to look more and more shortsighted.

The incentives extended to foreign investors were aimed at producing a mix of passenger cars and transport vehicles. Trucks and buses would provide the foundation for moving goods and people. As the economy grew, and as more Brazilians prospered—so the reasoning went—the huge and barely tapped consumer market for passenger cars would accelerate the growth of the automotive industry. In the 1980s, some economists predicted that Brazil would be the world's largest market for passenger cars by the year 2000. Brazil certainly has a *potentially* huge market. In the mid-1980s, U.S. and European manufacturers competed fiercely in markets that already had ratios of nearly one car for every two persons. Brazil, on the other hand, had but one car for every fourteen inhabitants. The big obstacle to exploiting this potential market was that so few Brazilians had the resources to even contemplate purchasing an automobile.

Without the generation of electric power, the impressive industrialization after 1967 would not have been possible. As they did with steel and automobiles, the technocrats built on the foundations laid by Kubitschek in the 1950s. The control and distribution of such goods and services as gas, water, electricity, and telecommunications are vital to a modern economy, and they are often matters of heated public debate. In the United States, a mix of public and private investment controls the so-called public utilities. Through the control of rates, the state and federal governments both promote and limit the private profits generated from the exploitation of "public goods." The Brazilians followed a pattern somewhat like that in the United States by setting rates, but economic nationalism and government policies have created a public utilities sector that is dominated by public capital.

Until World War II, foreign companies—primarily British, U.S., and Canadian—built and controlled the systems for producing and distributing gas, water, electricity, and information. A combination of rising nationalism and low rates (set by the government) gradually led to the replacement of foreign-controlled companies with state-controlled corporations. Nationalists on both the political right and left concurred in their desire to have public utilities under national control. Although the pattern historically may have been to shift from private, foreign capital to public, domestic capital in the control of utilities, the nationalism of the Brazilians has a strong parallel with the situation in the United States. Over the last decade, for example, the

United States government has repeatedly voided the sale of important telecommunications firms, citing "national security interests" as the rationale.

Under left-leaning administrations and the right-wing generals, Brazil moved gradually toward public control of key utilities. Beginning in the 1950s with the Vargas administration, federal, state, and municipal governments in Brazil took control of the production and distribution of energy, water, and communications. In the telecommunications industry this has, no doubt, led to lower quality of products and service. The Brazilians have not been able to compete with the sophisticated multinationals of the industrial nations, and protectionism has hurt the Brazilian consumer. Telephone equipment and services, for example, are far behind that in the industrial nations.

In the electric power sector the Brazilians have achieved incredible successes. State-controlled electric companies (most prominently CEMIG in Minas Gerais) have rapidly expanded the power grid, especially in the densely populated South and Southeast. The Brazilians have constructed one of the largest hydroelectric power systems in the world. The massive Itaipú Dam on the Paraná River at the Paraguayan border, built primarily by Brazilian companies, is one of the largest hydroelectric dams in the world. From the late sixties to the early eighties, electrical capacity expanded from 10 million kilowatts to more than 135 million, an extraordinary increase. The management of the state enterprises, in general, has been excellent, and the electric power industry has produced many of Brazil's most capable technocrats. In addition, rather than simply responding to the demand for power, management in the industry has actively promoted industrialization as a means of promoting their own future growth.

The strong growth of these three industries—iron and steel, automotive, and electric power—has been critical to the extraordinary industrial expansion that accelerated in the 1950s and exploded in the 1970s. They have both driven and been driven by each other's growth. All have grown with a varying mix of government intervention, foreign investment, and domestic entrepreneurship. Both civilian and military governments recognized that if Brazil were to become an industrial power these critical industries had to be nurtured and developed. Without them, Brazil could not hope to enter the exclusive club of industrialized nations.

These three industries also provided the foundations for the growth of another industry that the military regime zealously promoted and saw as the key to power in the world—defense. Building a strong defense industry served a variety of purposes for the armed forces. Greater self-sufficiency in the production of arms and equipment would decrease Brazil's long-standing and nearly total dependence on foreign suppliers, primarily the United States. Strengthening the autonomy and power of the armed forces would enhance the role of Brazil in the international community, especially among developing nations. A strong defense industry would also reinforce the general move toward economic growth and industrial power. What was good for the defense industry would be good for the national economy.

Although the Brazilian military throughout this century has tried to foster the creation of the defense industry and defense-related industrial growth, Brazil depended almost entirely on foreign imports of arms and supplies well into the 1960s. Spurred on by the euphoria of the regime's "economic miracle," and irked by the unwillingness of the United States to share technology in the Vietnam era, the military promoted the defense industry in the 1970s with spectacular results. In the mid-sixties, Brazil's production of defense hardware was insignificant. By the 1980s, Brazil had become the world's fifth leading exporter of armaments.

As in the United States, privately owned corporations produce the majority of defense hardware, but the federal government is the dominating force in the expansion and contractions in the industry. As early as 1969, the regime created Embraer to design and manufacture aircraft. In 1975, the regime created Imbel (Indústria de Material Bélico do Exército), which acted as a sort of superagency for planning, promoting, coordinating, producing, and exporting armaments. The military encouraged multinational corporations to invest in defense industries, and it created a variety of institutes and schools to train engineers and technicians for defense work, and to promote research and development.

The results were nothing short of phenomenal. By the late 1980s, more than 150,000 people worked in more than 650 firms in the defense industry, and sales of armaments to foreign buyers passed $3 billion. Almost entirely dependent on foreign suppliers in the 1960s, the Brazilian military probably imported less than 20 percent of its arms and supplies by the late 1980s. By some estimates, Brazil was the only Latin American country with the capability of going to war without having to import arms. Clearly, the creation of a defense industry had strengthened the overall process of industrialization and simultaneously strengthened the power of the armed forces.

The Brazilians achieved this impressive success through a careful strategy. Combining government support with both domestic and multinational investment, they developed relatively unsophisticated products that were aimed at specialized markets, especially in the developing world. Brazilian manufacturers wisely opted not to compete with the major industrial powers in high-technology markets. Instead, they created simple, sturdy, reliable products; subsidized their export; customized the products for customers; provided ample on-site technical assistance; and attached no strings to the sales. This strategy opened up markets all over the Third World, from Colombia to Iraq and China. The Iraq-Iran War was a powerful stimulus to the Brazilian defense industry, because the Iraqis purchased large numbers of armed vehicles and small arms. In exchange, Iraq became Brazil's major supplier of oil.

Although Brazil produces a very wide variety of defense hardware, missiles, aircraft, armored vehicles, and ships account for the bulk of exports. Four companies (known as the "four sisters") have dominated production, and as is true in other Brazilian industries, production is concentrated in the states of São Paulo, Rio de Janeiro, Minas Gerais, and Rio Grande do Sul. In the mid-1980s, Avibrás, a firm that produces missile systems, satellite communications systems, and air-to-ground defense systems, accounted for 40 percent of all arms sales. Engesa, one of the world's largest exporters of armored vehicles, accounted for another 30 percent. Embraer's aircraft sales had a 20 percent share, and Bernardini, a manufacturer of tanks, followed with 5 percent. Much like the Israelis, the Brazilians created their defense industry to supply themselves, knowing that the industry could not survive without export sales. Brazilian companies export approximately 90 percent of their output. Brazil and Israel together produced about half the arms exported in the world in the late 1980s.

The Brazilians have stimulated the growth of the defense industry through licensing agreements and joint ventures. Embraer has made arrangements to build aircraft in the United States and Great Britain in cooperation with companies in both countries. Avibrás has entered into a joint venture with the Chinese to build satellites, missiles, and aerospace equipment. Engesa (Engenheiros Especializados) has arranged to build tanks in Saudi Arabia. Prior to the Gulf crisis of 1989, Brazilian companies were working with Iraq on missile systems and nuclear weapons development. The Brazilians have created their own multinational corporations in the defense industry.

Despite the extraordinary success of the defense industry in the past twenty years, this development strategy has brought with it a number of problems. On a purely economic level, the boom in arms sales has leveled off and will most likely slacken in the next few years with the disintegration of the Soviet Union and the end of the arms race. As in the United States, in Brazil the defense industry, which is heavily dependent on exports, will have to retool and redirect some of its research and production toward civilian markets. In the 1990s, many of the defense industries have been privatized or have faced financial collapse with the end of government subsidies and protection. The Brazilian industry also will have to continue to develop increasingly sophisticated products to remain competitive.

More importantly, the promotion of the defense industry has created serious distortions in the national economy and created a military-industrial complex that civilian governments have seriously criticized. Much like the debate in the United States, many Brazilians have questioned the role of defense in research and development, as well as the morality of pouring scarce resources into the production of weapons of destruction. In short, would not this money be better spent in the creation of jobs, goods, and services for the millions of needy Brazilians, and not in the creation of weapons that could wreak, and have wreaked, destruction across the Third World? As Brazil enters the twenty-first century, the defense-industry strategy of the military regime, no doubt, will be seriously challenged, if not dismantled.

FROM MIRACLE TO DEBACLE

The technocrats and generals produced truly astonishing economic growth and built the largest industrial park in the developing world. Yet their growth strategy was also seriously flawed. Surprising shifts in the international economy, as well as poor policy decisions within the regime, brought an end to the apparent "miracle" of the seventies, leaving Brazil a legacy of economic disaster by the end of the 1980s. The military regime moved the country's economy forward faster than Juscelino Kubitschek's famous "fifty years in five," but only at a very high price.

Two major international shocks triggered the demise of the technocratic "miracle," and of the regime itself: the oil shocks in 1973 and 1979, and the double-digit interest rates of the late 1970s. Rising oil prices dealt Brazilian economic planners a staggering blow. They had built an economy on cheap, imported petroleum. In fact, the automotive industry's expansion had been a linchpin of the rapid industrialization. In the process, Brazil had become the largest importer of oil in the developing world, relying on imports from the Middle East for 80 percent of its needs. In 1973, the cost of importing oil quadrupled within months, draining off valuable capital, creating a balance-of-payments problem, and threatening to bring a halt to growth. (The oil price hikes alone added $2 billion to Brazil's import bill in 1974.)

Led by the widely respected banker, Mário Henrique Simonsen, the economic team of Ernesto Geisel's new administration took control of planning in the immediate aftermath of the first oil shock. Refusing to scale back and accept lower growth rates, Geisel's economic team opted to borrow their way out of the balance-of-payments crisis and to continue financing growth. Ironically, the oil crisis facilitated their borrowing strategy. Western banks awash in the vast "petrodollars" accumulated by the oil-exporting nations swept across the world in the mid-1970s, wildly handing out loans to virtually anyone. With its impressive economic track record over the previous decade, Brazil offered the bankers enormous possibilities for "recycling" their petrodollars.

The Geisel administration also launched two massive and expensive programs to diminish Brazil's reliance on imported oil using nuclear power and alcohol. Both programs accelerated the growth of government intervention in the economy through the creation of state enterprises. In the short term, both programs siphoned off valuable and scarce resources rather than diminishing the impact of high oil prices. To many, the alcohol program seemed a visionary and bold alternative to petroleum dependence, and the program received a great deal of favorable international attention.

The nuclear program, in contrast, generated widespread international criticism and condemnation.

The Brazilian military had long-standing ambitions to develop a nuclear industry for both military and economic reasons. They recognized that the establishment of a domestic industry would put them into a very exclusive international club. In both economic and military terms, the nuclear industry could contribute to the drive to make Brazil a power in the world by the end of the century. Spurred on by the oil crisis, the regime signed a sweeping accord with West Germany in 1974 that would give the Germans access to Brazil's uranium reserves and the Brazilians access to German nuclear technology.

Although Westinghouse had been constructing a nuclear power plant at Angra dos Reis (on the coast between Rio de Janeiro and São Paulo) for several years, the United States government and U.S. corporations refused to give the Brazilians full access to key technology. The West Germans, in contrast, offered technology and technical assistance for every step of the nuclear fuel cycle—from prospecting and processing uranium, to designing reactors, to reprocessing spent fuel. Reprocessing aroused the greatest opposition from the United States, since it would give Brazil the capacity to produce weapons-grade plutonium. In effect, the accord gave the Brazilian military regime access to nuclear technology that would have both commercial and military uses.

While the nuclear program seems fraught with technical, economic, and moral problems, the alcohol program (known as Pro-Alcohol) appeared to be a very bold, yet reasonable option. Brazil produced more sugar cane than any other country in the world (except India), and the program addressed two important needs with a single solution. The impoverished Northeast and the perpetually troubled sugar industry produced enormous amounts of a raw material that could be converted into alcohol as a substitute for high-priced imported oil. Pro-Alcohol proposed to develop the technology to convert the massive cane production of the Northeast into fuel for automobiles. The program would stimulate domestic industry and reduce the amount of valuable currency spent on foreign oil.

Despite its otherwise abundant resources, Brazil does not appear to be blessed with large petroleum reserves. Acutely aware of the importance of petroleum for modern economies, the Brazilians nationalized the oil industry as early as 1953 after a long and intense campaign by nationalists. Until very recently, a state enterprise, Petróleos Brasileiros (Petrobrás), had a monopoly on oil prospecting and production. Formed even before significant oil reserves had been discovered, Petrobrás has gradually located and developed oilfields, principally off the coast of Bahia in the Northeast, and in Rio de Janeiro in the Southeast. In recent years, the company has begun to explore the vast Amazonian basin for oil. Results have been modest. Optimistic projections call for production of 50,000 barrels a day in the Amazonian operations by 1996. In 1994, Petrobrás produced about 700,000 barrels of oil a day, about 60 percent of what Brazil consumed.

By the late 1970s, Petrobrás had become the largest corporation based in Latin America (with 47,000 employees and more than $18 billion in revenues), and Pro-Alcohol became one of its most important projects. With the price of oil near $40 a barrel in 1974, the company poured millions of dollars into the development of refineries in the Northeast to produce alcohol for automobiles that would run on 100-percent-hydrated alcohol. By 1990, more than 4 million Brazilians drove alcohol-powered vehicles. (In 1985, 90 percent of all automobiles coming off Brazilian assembly lines were alcohol-powered. By 1995, that number had fallen drastically to 5 percent.) The costs of producing alcohol were high, and the government had to subsidize the cost to offer consumers alcohol at a price level equivalent to gasoline. As the price of oil dropped in the 1980s, the amount of subsidies increased rapidly and began to bleed Petrobrás of revenues. The company found itself in a terrible bind. If it reduced subsidies, it choked off the Pro-Alcohol program that had already absorbed billions of dollars of investment. If it did not, the company would lose money. In the early 1990s, the government continued the subsidy program, and Petrobrás racked up huge losses.

Like the rest of the economic strategy, both the nuclear and the alcohol programs required huge amounts of capital investment, which made the Geisel administration ever more dependent on external borrowing. As a result of this strategy of borrowing the country out of the crisis and into continued growth, Brazil's foreign debt escalated rapidly. In 1974, Brazil owed about $10 billion to foreign creditors, a figure that doubled by the end of 1975. When the Figueiredo administration took office in 1979, and the next oil shock struck, Brazil's external debt had passed $40 billion and continued its rapid rise. By the end of the Figueiredo administration (and the military regime) in 1985, the external debt had passed $100 billion.

Nearly every major sector of the economy relied on foreign borrowing. State-controlled enterprises and banks, as well as private corporations and banks, depended heavily on a continuing influx of foreign capital. The state-controlled banks and industrial enterprises channeled borrowed capital into key sectors of the economy, stimulating production, increasing exports, and generating revenue to keep the system going. As Albert Fishlow has pointed out, this outward-oriented growth that increasingly integrated Brazil into the world economy was asymmetrical. That is, Brazil's share of debt grew faster than its share of trade. As long as the Brazilians could keep the foreign investment flowing in, and the exports flowing out, the system worked. Should imports rise too quickly, or should interest payments on the loans rise too rapidly, Brazil's trade surplus would diminish and it would face another balance-of-payments crisis.

Three powerful shocks to the world trading system in the late seventies and early eighties threatened the growth strategy of the military regime's technocrats. Ironically, the architect of the earlier "miracle," Antônio Delfim Neto, once again commanded the regime's economic policy, this time for President João Figueiredo. It was Delfim Neto who struggled to respond to the triple threat of a new oil shock, escalating interest rates, and recession. Rather than taking a cautious strategy to soften the blows to the Brazilian economy, Delfim tried to combat these three shocks with more borrowing and more exports. The strategy failed.

In fairness to Delfim Neto and his team, they did have less room to maneuver than they had during the first oil shock. The debt-financed growth policy of the seventies had created a huge external-debt burden, and the country continued to import large quantities of oil. On the other hand, the planners continued to rely on export-led growth financed by foreign loans despite the clear signs of recession and escalating interest rates. The exceptionally high interest rates increased the burden of servicing the debt as interest payments quickly passed $5 billion a year. Expensive oil added to the balance-of-payments problems. As demand for Brazil's exports lagged in 1981 and 1982, the crisis came to a head.

Much like Mexico, Brazil faced a liquidity crisis in 1982. Interest payments on the debt and the cost of imports (especially oil) surpassed the value of exports, and Brazil found itself unable to meet its international financial obligations. International banks, which had been lending Brazil close to $1 billion a month in 1981, shut down the cash flow. Unable to get new loans, the Brazilians now had to cut back internally. Industrial production declined, unemployment rose, a severe recession set in, and a decade of hard times began. The miraculous expansion turned into a nightmarish recession.

THE PRICE OF PROGRESS

As if the enormous difficulties created by the debt crisis were not enough, the economic policies of the military regime created a variety of problems that it would bequeath to the civilian politicians in 1985. Two decades of extensive government intervention had made the economy the most statist in the non-communist world. After some initial reductions in the inflation rate, inflation exploded in the 1980s, surpassing 100 percent for the first time ever. The growth model, followed by the military regime, also intensified the economic inequalities within Brazilian society. The richest 10 percent of the nation had control of more wealth in 1980 than in 1960, and the poorest 40 percent had less. Finally, the regime's ambitious drive to develop the vast and sparsely settled interior launched an era of devastation and ecological havoc in the Amazon basin.

The right-wing military regime that proclaimed itself capitalist and anti-communist left a legacy of state intervention unmatched outside the socialist bloc. By the mid-1980s, the state owned the twenty-five largest nonfinancial corporations in Brazil. In the financial sector, government banks accounted for nearly 40 percent of the loans held by the nation's 50 largest banks. The state-owned Banco do Brasil alone held nearly a quarter of all funds on deposit in the same banks. According to one estimate, in 1981 the nonfinancial state enterprises spent $73.5 billion, an amount equal to nearly 30 percent of the nation's Gross Domestic Product. The state enterprises had borrowed heavily overseas and at the beginning of the 1980s held about half of the country's total foreign debt.

The bulk of the state enterprises were clustered in the sectors of the economy that supply crucial goods or services to key industries: iron and steel, electricity, banking, nuclear power, and oil, to name a few. The electric power industry alone consumed 10 percent of all domestic investment in the late 1970s. Although a number of the state enterprises have performed well and made vital contributions to national development, the state sector has become too bloated, expensive, and bureaucratized. Despite "islands of excellence," many of the state enterprises are unnecessary and have become a drain on the government's limited resources as they continue to lose money and borrow heavily. The creation of more than 175 state enterprises has also been accompanied by a morass of laws and regulations that have hindered rather than stimulated economic growth.

Government efforts to control inflation have also seriously damaged the national economy—arguably doing more to perpetuate inflation than to eliminate it. Brazil has suffered from chronic inflation throughout this century. In the postwar period, the inflation rate has generally been around 20 percent on an annual basis. The economic and political chaos of the early sixties pushed the rate to near 100 percent in early 1964, which was a major factor contributing to the fall of Goulart. Under the austerity measures of Campos and Bulhões, and the growth strategy of Delfim Neto, inflation came down, but rarely below 20 percent a year. In 1980, as Delfim Neto's economic strategy fell apart, inflation exploded and passed 100 percent. The rate hovered around 100 percent in 1981 and 1982. As President Figueiredo lost control of the process of *abertura* in 1984-85, he also lost control of the economy, and the civilian government inherited an economy threatened by rampant inflation running at an annual rate of 200 percent.

One of the major contributors to the chronic and continually rising inflation was one of the most interventionist measures of the military regime—indexation. Although indexation (or "monetary correction" as it is called in Brazil) had been used in small ways as far back as the early fifties in Brazil, it was the military regime that expanded its use and made it a central fixture in the economy. The idea behind indexing was simple enough: to adjust certain financial instruments to ensure that investments would stay ahead of the eroding effects of inflation. This would encourage domestic savings and would mobilize capital. Historically, Brazil has had a very low savings rate and, consequently, has been unable to mobilize the domestic capital to finance economic growth. The Japanese, in contrast, have traditionally had a very high internal savings rate that has played a key role in mobilizing the capital to finance their extraordinary economic growth.

Indexing began in a modest fashion under Campos, and then gradually became a central tool in the hands of economic planners. In 1964, the regime created new government bonds called ORTNs (Readjustable National Treasury Obligations), with the principal adjusted monthly based on a price index. The ORTNs eventually became the central "unit of account" used as the baseline for all financial accounting. In effect, Brazil had two "units of account," the ORTN and the *cruzeiro*. This caused distortions in the economy as financiers speculated moving their funds according to fluctuations in both units. In a sense, Brazil operated with two currencies. Although Campos envisioned "monetary correction" as a temporary measure, it was soon extended to mortgages, insurance, passbook savings deposits, and a wide variety of financial instruments. The government also periodically readjusted prices and wages, although they were not tied specifically to the same formulas as the financial markets. By the mid-seventies, the government had intervened into virtually every sector of the economy through some form of indexation.

On the positive side, it does appear that indexation helped stimulate savings and mobilize capital in the form of government bonds, savings accounts, and housing bonds. For reasons that are still hotly debated, indexation did not mobilize capital in the industrial sector that turned to government credit and foreign borrowing to satisfy its capital needs. More importantly, indexation has saddled Brazil with two pernicious problems: speculation and built-in inflation. The uncertainties created by the constant shifts in artificial readjustments in all sectors of the economy has been a deterrent to long-term investment and has stimulated short-term speculation on the financial markets. Much as they did during the "junk bond" decade of the eighties on Wall Street, a lot of investors made money without producing anything tangible. Finally, indexation has created a vicious cycle that both military and civilian leaders have found difficult to break. The government has tried to diminish the investors' risks by absorbing the risk through indexation, but it has done this by building inflation into the system. The legacy of this strategy has been the runaway inflation of the past few years and an inability to break out of the cycle.

Another disturbing and controversial feature of the Brazilian growth model has been increasing inequity in income distribution. Interpreting the statistical data on income inequalities has become something of a cottage industry among economists and has provoked considerable disagreement. If there is a consensus on the data, it is that (in relative terms) the rich got richer and the poor got poorer in Brazil in the period from 1960 to 1980. According to the World Bank, Brazil now has the most inequitable distribution of wealth of any country in the world. In 1960, the richest 20 percent of the country's economically active population accrued 55 percent of the nation's income, while the poorest 40 percent earned just 11 percent. By 1980, the richest 20 percent had increased their share to 63 percent, and the share of the poorest had declined to just under 10 percent.

Despite increasing income inequities in relative terms, in absolute terms Brazilians benefitted from a rise in real income. Furthermore, economic growth created millions of new jobs, many of them skilled and well paid. In absolute terms, the nation made substantial economic progress. Unfortunately, the more educated and more skilled workers received most of the new wealth generated by growth. The economic pie grew, but very few of the benefits "trickled down" to the poorest Brazilians. The gap between rich and poor widened, a dangerous trend in any country, but especially one in which the poor make up more than half of the population. The "miracle" generated impressive economic growth, but at what may prove to be an unacceptably high social cost.

Perhaps the most enduring and far-reaching legacy of the military regime's economic strategy was the opening of the Amazon basin. Although Amazonia comprises more than 40 percent of Brazil's land-mass, in the 1960s less than 4 percent of the Brazilian population resided there, mostly in the cities of Manaus in the western Amazon and Belém on the Atlantic coast. Except for the shortlived rubber boom at the turn of the century, when Brazilian rubber became essential to the industrial nations, the region had largely been ignored by successive Brazilian governments and investors, both foreign and domestic.

In the late sixties, the military regime made the dramatic, and apparently irreversible, decision to integrate Amazonia into the national economy and to exploit its vast and untapped resources. From a geopolitical perspective, they believed that Brazil had to establish a strong presence in the region, especially on the border regions. Nationalists within the military have long believed that neighboring nations and the industrial powers wish to wrest control of Amazonian riches from Brazilians. In the words of the military, they had "to integrate so as not to surrender" *(integrar para não entregar)*. From an economic perspective, the generals longed to exploit the enormous mineral and vegetable wealth of Amazonia. Finally, the vast and sparsely occupied lands of the Amazon seemed to offer the perfect solution to the old conflicts over land in the densely populated Northeast. In the famous phrase of the era, the regime hoped to unite "men without land to land without men."

The military set up "Operation Amazonia" in 1966 to promote the exploitation of the region. A new government agency, the Superintendency for the Development of Amazonia (SUDAM) created fiscal incentives to attract investment in agriculture, cattle ranching, and industry. Investors were granted tax exemptions, subsidies, and credits, and the newly created Bank of Amazonia provided loans. In 1970, President Médici initiated the Plan for National Integration (PIN), embarking on an ambitious road-building program aimed at linking up the vast interior with the populated coastal regions. The following year, the government created its Land Redistribution Program (PROTERRA), which was designed to promote modernized agriculture in the Amazon, an objective that translated largely into promoting agribusiness.

The road-building program became the key to opening up Amazonia. Over the next two decades, the government planned and built roads to crisscross the Amazon from the western borders to the Atlantic coast, and from the northern jungles to the densely populated industrial Southeast. The Trans-Amazon Highway alone would run 2,700 kilometers from the Peruvian border to Recife and João Pessoa on the northeastern coast. One major north-south artery would connect Santarém (midway up the Amazon River) with São Paulo via Cuiabá (the capital of Mato Grosso), and another would link Belém and the Southeast via Brasília. Over the next two decades, Brazilian construction companies cleared rain forest and laid down thousands of kilometers of road through Amazonia. Three of the country's largest corporations— Camargo Corrêa, Andrade Gutiérrez, and Mendes Júnior—benefitted handsomely from government road-building contracts.

These dirt and asphalt highways became the Brazilian equivalent of trails across the American West in the nineteenth century. Streams of immigrants flowed into Amazonia by the bus- and truckload in the wake of the road-building crews. The government claimed the right-of-way on either side of the new highways, and created the Institute for Agrarian Reform and Colonization (INCRA), ostensibly to settle up to 5 million people along these roads by 1980. Tens of thousands of peasants moved down the new highways in the 1970s and 1980s, staking a claim and then burning off the rain forest to clear the land for farming and cattle ranching.

In the 1960s, the movement of migrants flowed mainly into the eastern Amazon basin along the newly paved Belém-Brasília Highway (BR-153). Some 300,000 to 400,000 people streamed into northern Goiás and eastern Pará, staking claims and clearing the forest. In the 1970s and 1980s, a flood of migrants swept across the western Amazon, following the opening and paving of BR-364 from Mato Grosso into the state of Rondônia on the Bolivian border. Between 1970 and 1985, the population of Rondônia boomed from under 100,000 to over 700,000. Western Amazonia in the 1980s became a contemporary version of the Old West in the nineteenth century, with its chaotic frenzy of immigration, colonization, and frontier "justice."

The dream of land has proven elusive for the hundreds of thousands of poor peasants who became Brazil's newest pioneers. Poor soils, the difficulties of transporting crops to markets, malaria, and the hostility of a new breed of large landowners—the cattle ranchers—dashed their hopes of building a new life in the Amazonian forests. The cycle became all too familiar. Migrants burned the forest, failed miserably at planting in the poor soils, starved and took ill, and then abandoned the land, often to the cattle ranchers. With few jobs available to them on the ranches, they often migrated into the growing slums of Manaus, Belém, or the newer cities in Pará and Rondônia. The dream of turning the vast "land without men" into farmland became a nightmare.

As the colonization of Amazonia went awry, the plan to develop its resources also became a human and environmental disaster. Most of the thousands of square miles of deforested land became pasture for large cattle ranchers who took advantage of the tax credits and fiscal incentives offered by SUDAM. Most of the ranchers made their profits off the tax breaks and had little incentive to turn a profit from beef production. Multinational corporations (like Volkswagen) and wealthy businessmen in São Paulo invested in cattle-ranching ventures, profiting handsomely from unproductive enterprises. As the ranchers burned larger stretches of forest, they came into open and often violent conflict with poor peasants. Poor immigrants soon found themselves replaying the violent land conflicts they had fled in the Northeast and Southeast. Government fiscal policies, in effect, encouraged deforestation and violence.

(Until very recently, the Brazilian government has shown little interest in environmental protection. In the early 1970s, planning minister João Paulo Velloso responded to criticism of a newly approved paper mill with an attitude that was fairly typical of technocrats and generals at the time. When opponents showed that it would pollute Amazonian waterways, Velloso conceded that it would, and retorted, "Why not? We have a lot left to pollute.")

Mining has been the most capital-intensive sector of the Amazonian development program. The systematic surveying of the region in the 1960s and 1970s revealed vast mineral deposits of gold, uranium, bauxite, and iron ore, just to name a few key minerals. Some estimates place the value of the region's minerals at close to $2 trillion. The lure of these rich resources has attracted investors from all over the globe. The Brazilian government, Brazilian entrepreneurs, and multinational corporations have poured hundreds of millions of dollars into mining projects across the region over the past twenty years.

The biggest project of them all is Carajás, an enormous series of mining operations centered in the lowlying Carajás mountain range about 350 miles south of Belém. Discovered by accident in the late sixties, the mountains hold the world's largest iron ore deposits, possibly as much as 20 *billion* tons of rich ore. With funds from Europe, Japan, the United States, and international lending agencies, the Brazilian government and CVRD have constructed a huge iron mine, smelters, and large hydroelectric dams to provide power for operations. In addition to cutting roads into a region that was once inaccessible to all except by canoe and helicopter, they built a 500-mile long railway to São Luis, the capital of Maranhão on the Atlantic coast. The CVRD built ten new cities just to house workers.

The dams have flooded thousands of square miles of rain forest and displaced previously undisturbed Indian tribes. Immigrants flocked to the area looking for work and land, further devastating traditional Indian society. The smelters have been fueled by charcoal rather than by more expensive imported coal. The surrounding forests have been leveled to supply the charcoal to make pig iron. In 1990 alone, loggers chopped down more than a million trees around Carajás. The dense smoke emitted by the smelters reproduces, on a smaller scale, the same effects as the burning of huge tracts of forest in Rondônia. Many scientists believe that the fires and smoke from Amazonia aggravate the so-called Greenhouse Effect, gradually raising the Earth's temperature.

The unprecedented rise in the price of gold in the late 1970s triggered a gold rush in the Amazon that has also had profound environmental impact. With few exceptions, gold in Brazil is located in alluvial deposits and not in deep-seated lodes or veins. Throughout the 1980s, tens of thousands of prospectors (*garimpeiros* in Portuguese) swarmed across the rivers and streams of Amazonia, panning and dredging for ore. The influx of thousands of *garimpeiros* in the remote province of Roraima on the Venezuelan border has nearly destroyed the Yanomami Indians and their stone-age culture. The production of the *garimpeiros* is virtually impossible to track and goes largely to the black market. Well-informed observers, however, believe that the *garimpeiros* mine nearly ten times the officially recorded annual production of eight tons. If this is true, only Russia and South Africa produce more gold than Brazil.

The most spectacular discovery of the gold rush came in eastern Pará, to the east of the Carajás mountains at a place known as Serra Pelada ("bald mountain range"). In late 1979, a cattle rancher discovered gold nuggets at Serra Pelada, an event that touched off a small-scale gold rush. By mid-1980, some 30,000 people had come to Serra Pelada hoping to strike it rich. The federal government, which had first taken an interest in the region during the Araguaia guerrilla insurgency in the early seventies, soon took control of the entire area. In effect, the military ran the Serra Pelada mining operation. At its height in 1983, possibly as many as 100,000 people lived around Serra Pelada. Close to 50,000 men labored in an enormous open-pit mine and produced one metric ton of gold ore per month using rudimentary technology.

Today, Serra Pelada has been abandoned, and the crater has become a small lake. The *garimpeiros* have moved on to other promising areas, most prominently, Roraima. The legacy of *garimpagem*, however, is enduring. In addition to the deforestation and destruction of traditional Indian communities, gold mining continues to poison Brazilians. Using an ancient technique, the *garimpeiros* separate gold ore and purify it with mercury. The runoff of mercury into the Amazonian water system will have long-term consequences. Studies have already shown dangerously high and toxic levels of mercury in the blood of peasants and Indians in gold mining regions. The mercury continues to run off into thousands of rivers and streams throughout the Amazon basin as *garimpeiros* continue their search for the precious ore. Long after the gold is gone, this legacy of the gold rush will haunt Brazil.

A DECADE OF FAILED PLANS

The legacy bequeathed by the military regime to civilian politicians in 1985 was not just extraordinary economic growth, but a peculiar kind of industrialization with some profoundly disturbing economic, social, and environmental flaws. The Sarney administration, with its indecision and ineptitude, did little to confront the crisis created by the military regime. Sarney's major effort to confront the crisis came in February 1986 with the Cruzado Plan, which was geared at bringing triple-digit inflation to an immediate halt. This "heterodox" program froze prices, froze wages after a substantial pay raise, fixed the exchange rate, and created a new currency, the *cruzado* (equal to 1,000 of the old currency, the *cruzeiro*). Inflation, which had been 22 percent in February alone, dropped to zero in March, April, May, and June. Millions of Brazilians became self-appointed "price inspectors" (*fiscais*) of the president to enforce the freeze. The wage hike, however, sent consumers on a spending spree that overheated an already hot economy. For a few brief months, Sarney was the most popular politician in Brazil and enjoyed massive public support.

The Cruzado Plan became such an important political success that Sarney ignored advice from his economic team to readjust prices gradually. Instead, in a blatantly political move, prices remained frozen until after the November 1986 congressional elections. One week after the elections, the government announced Cruzado II, and hiked prices dramatically in key areas. The results were immediate and dramatic. Inflation revived and within months was again averaging more than 20 percent per month (or more than 2,000 percent per annum). Sarney never recovered politically from this debacle. The Brazilian economy stumbled along without strong and decisive leadership for the next three years. Finance ministers came and went, as did their failed economic plans.

In contrast to the indecision and drift of the Sarney presidency, Fernando Collor de Mello immediately mounted a frontal assault on the economic crisis as well as the economic policies of the previous fifty years. The day after his inauguration in March 1990, he unveiled the New Brazil Plan, the most sweeping economic austerity program in the nation's history. In the long term, Collor declared that he wanted to move Brazil from the Third World into the First World. In the short term, he was trying to accomplish what two previous presidents and a series of austerity programs failed to do in the last decade: stop quadruple-digit inflation and get the economy back on track toward steady growth. After five years of political drift and a decade of economic crisis, everyone expected the new administration to take dramatic steps. Few anticipated the severity of the measures in store.

The principal target of the plan was the runaway inflation that has plagued the country for the last decade, reaching four digits by 1988. Rather than simply freeze prices and wages, the new government took the extraordinary step of declaring an eighteen-month freeze on all savings accounts in excess of 50,000 *cruzados novos* (about $1,200 at the official rate of exchange). This measure tied up about $80 billion, or about 80 percent of the currency in circulation (although it only affected about 10 percent of the population). Eventually, the frozen funds were to be redeemed in *cruzeiros* (the new currency and the fourth in four years) in twelve installments, without adjustment for inflation. Several commentators called this the biggest liquidity squeeze in Brazilian history.

In addition to the severely restrictive monetary policy, Collor imposed a thirty-day wage-and-price freeze and announced drastic measures to cut government spending, lay off 360,000 government employees, and sell off state-controlled enterprises. Collor wanted to "shrink" the state apparatus and get the government out of the business of economic intervention. In the drive to open up the economy, the government began phasing out tariffs, many state subsidies and fiscal incentives were suspended, and the exchange rate was allowed to float.

With a foreign debt of $114 billion (the largest in the world after the United States), a domestic debt of similar proportions, and a public-sector deficit of $25 billion, Collor entered office facing enormous challenges. He initiated changes more sweeping than any other political figure in twentieth-century Brazil. The controversial centerpiece of the plan, however, the freezing of bank assets, had mixed results. Perhaps no other figure than a president elected with the strong support of the right could have gotten away with such a profoundly interventionist move. Unwisely, Collor initially promised to "liquidate" inflation during his first 100 days in office. When the rate for April was 3-4 percent and then 10-12 percent in May, June, and July, the administration had to do some backpeddling to explain how the rates were "acceptable" and why expecting zero inflation, for the moment, was unrealistic. Inflation was in the range of 20 percent per month two years after Collor took office. Although clearly disappointing to those who expected the shock treatment to end inflation, the figures looked good in comparison with the 84 percent rate in February 1990, and the 2,700 percent for the twelve months prior to Collor's inauguration.

Collor's efforts to reduce the government payroll to save $3-5 billion also ran into problems. Sixty to seventy percent of the government's budget was going to administration, the majority to pay the salaries of more than 1.3 million federal public functionaries. To streamline the unwieldy bureaucracy, Collor reduced the number of cabinet ministries from 27 to 12 and ordered all government agencies to cut the payroll. Although officially nearly 160,000 employees were laid off, labor legislation has made it impossible to fire nearly a third of those. The government could send them home, but had to keep paying these "fired" employees their salaries. In addition, Collor was forced to back away from his announced mass firings to reach the goal of 360,000 by June 1990 when threatened with a nationwide strike by union leaders. Nevertheless, despite the partial fulfillment of the goal, the layoffs are the most impressive reduction in the history of the Brazilian civil service.

Efforts to cut the deficit by selling off state assets have also taken more time than anticipated. The machinery has slowly been put into place to privatize the majority of more than 250 state-controlled enterprises. Usiminas, the huge and profitable steel company, became the first enterprise to be auctioned off in the fall of 1991. The sale went forward after violent demonstrations by the left and labor unions in front of the Rio Stock Exchange prior to the scheduled auction. Despite fears of foreign buyouts, a consortium of Brazilian institutions purchased majority control of the company. The sale of state companies reverses a powerful nationalistic and interventionist pattern over the last half-century. In particular, the privatization of the major state-owned steel companies marks the end of an era. (As of July 1995, some thirty-four state enterprises had been auctioned off at a price tag of about $9 billion.)

The government has been unable to reverse the trend of deficit spending, perhaps the most serious problem facing Collor. Prior to the 1980s, the Brazilian government generally ran budget surpluses or moderate deficits. The combined pressures of the post-1982 financial squeeze and the return of mass politics initiated a decade of fiscal irresponsibility. In an era when capital flows slowed to a trickle, politicians at all levels of government racked up large deficits to promote their projects, and to employ their friends and allies. The federal government became the banker of last resort as municipalities passed on their financial problems to states, and the states to Brasília. The state-controlled banks refused to hold the line on spending and thus ended up holding huge amounts of unrecoverable loans.

Much like the United States Congress, the Brazilian legislature has been largely unwilling to make the profound and painful cuts necessary to balance the budget. Politicians recognize that a national consensus does not exist on how everyone should share the pain from the enormous budget cuts. Consequently, the Brazilian municipalities, states, and federal government continue to run deficits that feed inflation. Unlike the United States, Brazil has to face the deficit problem now. The U.S. government can continue to spend money it does not have as long as the Japanese and Europeans are willing to buy up treasury debt. Since 1982, foreign banks and investors have been unwilling to purchase Brazil's debt denying the government the resources to continue deficit spending.

Although the government budget deficits are viewed with some skepticism, the government's trade program has received widespread support in the international financial community. The announcement in early July 1990 that Brazil would move to drop virtually all trade barriers is, in a sense, the second part of the shock plan, and just as important as the internal adjustments. The dramatic effort to break with decades of protectionism is intended to modernize the economy and open it up to competition. The government hopes this will drive down prices and inflation, and raise industrial productivity. (A recent survey of the ten most industrialized countries in the developing world ranked

Brazil ninth in industrial productivity—just ahead of India.) Critics of the new trade policies fear a possible repeat of the Argentine experience of the late 1970s, when its economy was suddenly and drastically reopened, wreaking havoc on national industry.

Tariffs over the past few decades have sometimes run as high as 100 percent, which has helped to promote domestic industry, but in sectors that are often inefficient and heavily subsidized. The announcement of the new trade policies was especially welcome in the United States after nearly a decade of battles over protectionism, especially in the computer industry. In the early 1980s, the Brazilian congress passed an "informatics" law that attempted to foster a domestic computer industry by closing the national market to foreign computer hardware. The legislation has created a considerable industry in Brazil, especially for small computers, but has kept businesses and government from modernizing their operations with the latest technology from abroad. With the move away from protectionism, the United States has lifted punitive duties previously imposed on Brazilian goods and dropped the threat of additional sanctions.

Collor's austerity program and the new trade policies were major steps in the move to reopen negotiations on the foreign debt. From July 1989 to late 1991, Brazil did not make any payments on the debt. Interest payments resumed in late 1991, but the debt negotiations moved along in jumps and starts due to the intransigence of both sides and changes in the Brazilian negotiating team. Brazil needs about $15 billion a year to meet its obligations. This figure represented about 50 percent of the value of export earnings per year in the late 1980s and early 1990s.

Despite problems with payments, the International Monetary Fund, the U.S. Treasury Department, and foreign commercial banks finally worked out a restructuring of Brazil's external debt in 1994. The debt payment problems have taken a toll on the commercial banks and on Brazil. Citicorp, the largest private creditor, suffered a reduction of $400 million in 1990 pretax profits due to the interest payment moratorium. According to John Reed, Citicorp's chairman, Brazil's failure to honor its debts have cost it billions of dollars in short-term credits. The Brazilian debt has long been selling on the secondary financial markets at about 25 percent of face value, a sign of the skepticism and pessimism of the financial community.

The audacious move to open up one of the most closed free-market economies has been risky. Over the last decade, one of the bright spots in the Brazilian economy has been large trade surpluses, which reached levels as high as $19 billion in 1988. (Although the country ran $2-4 billion monthly deficits during much of 1995. In 1996, Brazil imported about $53 billion of foreign goods and exported about $48 billion of Brazilian products.) Over the last few years, only Japan and Korea have had healthier trade balances. Brazil needs to continue running surpluses over $10 billion to meet obligations on the debt, yet the reduction of tariffs could severely reduce the surplus with a rise in imports. In late 1994 and early 1995, Brazil consistently ran record high trade deficits, largely due to the demand for imports recently freed of costly tariffs.

The Iraqi invasion of Kuwait also aggravated the trade balance. Brazil imports about 50 percent of its oil (about 600,000 barrels a day), and around one-third of its imports came from Iraq and Kuwait prior to the invasion. During the last decade, Iraq was also a major importer of Brazilian goods, which ranged from frozen chicken to armaments. In 1989, Iraq sold about $1.5 billion worth of imports to Brazil. (Only the United States was a bigger purchaser of Iraqi goods.) The war in the Middle East both hurt Brazil's exports and raised its import bill.

The easing of the debt crisis has allowed Brazil to resume economic growth with a new influx of foreign capital. The crisis of the last decade, two moratoriums on debt payments, and the lack of a consistent set of economic policies have severely hampered the flow of fresh funds and led to capital flight nearing $30 billion. At a time when the nation desperately needed new investment, it became a major exporter of capital. Although the severe crisis of the eighties has passed, financial stability has not yet completely returned.

Despite the sweeping measures, Collor's economic plan failed to end the inflationary spiral. Inflation stabilized at around 10 percent per month by mid-1991, and then headed into the 20-40 percent per month range in 1992, 1993, and 1994. In 1993, the country suffered its worst inflation ever, passing 2,500 percent for the year. (The previous high was nearly 1,800 percent, in 1989.) Although technically the nation has not experienced hyperinflation, prices set on one consumer index at 100 in early 1986 had passed 3,000,000 by early 1991.

The cost of the anti-inflation program was a sharp recession and a rise in unemployment. During the first year of the Collor economic plan, one-third of Brazil's 500 largest businesses lost money. The economy shrunk in 1989, 1990, and 1991, and industrial output contracted during four of the five years between 1988 and 1991. In Greater São Paulo, the economic heartland of the nation, the number of unemployed doubled in the first six months of 1990, reaching 1 million. Real wages were down by 30 percent over the same period. Real GDP in 1991 had dropped to its 1986 level, and real GDP per capita had declined as well.

A parade of more than a dozen finance ministers, five drastic economic plans, and four changes of currency between 1985 and 1994 failed to pull the country out of its decade-long economic crisis. Brazilians tried all kinds of economic theories and plans, and none seemed to work. The Brazilian economic experience clearly does not fit very well into traditional economic theories and categories. Yet the biggest failure over the last decades has not been theoretical, but practical. No government has been able to apply its program and overcome the many obstacles presented by entrenched interest groups and institutions. Even the daring Collor Plan that struck directly at the powerful interests of the elites and middle classes has not dismantled many of the most important and most deeply entrenched programs and policies.

NEW HOPE FOR THE FUTURE

The election of Fernando Henrique Cardoso to the presidency and his first two years in office offer tremendous hope, and great expectations, for Brazilians. After nearly fifteen years of failed economic plans, weak political leadership, and scandals, Brazil seems to have been set back on the track to economic growth and stability by Cardoso. Despite his long-standing left-wing politics and credentials, Cardoso has clearly taken up the agenda of neo-liberalism (read conservative in U.S. politics): privatization, freer trade, less state intervention, and an end to economic nationalism.

Cardoso's successful Real Plan paved his path to the presidency. As finance minister in 1994, he instituted an economic plan that seems to have worked. The new currency (the *real*) has been strong and has not experienced the continual devaluations of early currencies. Inflation has been reduced to 1 percent per month (from July 1994 to December 1996), giving Brazilians their lowest inflation and most stable currency in four decades. Economic growth has been strong, and foreign investment has begun to flow into the country on a scale not seen since the "miracle" years of the 1970s ($9 billion in 1996).

Cardoso's economic program and its success, however, hinge on his ability to forge a political transformation, a task he seems to be accomplishing. During 1995 and 1996, Cardoso pushed fundamental constitutional reforms through congress, ending state monopolies on the petroleum industry and on telecommunications, and ending long-standing restrictions on foreign control of national resources and industry. Although moving much slower than Chile or Argentina, Brazil has clearly made a major shift in the direction of a more open economy with diminishing state control and intervention.

Brazil seems to have made a profound and irreversible shift, ending more than fifty years of statist intervention, economic nationalism, and protectionism. Yet, Cardoso faces several major challenges. Brazil has an incredibly inefficient and inequitable tax system, which not only hinders the growth of government revenues, but also promotes tax cheating. Like many Latin American nations, Brazil has never had an effective income tax. Only about 5 percent of tax revenues come from income taxes. As a result, the economy is permeated with all sorts of sales and transaction taxes that hinder economic growth, encourage underreporting of sales, and lead to all sorts of bureaucratic obstacles in the business community. Cardoso must push through a rational and efficient reform of the tax structure of Brazil.

On the expenditure side, he must continue to cut government spending to end the annual deficits that contribute to inflation and fiscal irresponsibility. This is perhaps his toughest task, since virtually all politicians see the state as the source of patronage for their political supporters. To cut spending is to battle over whose constituents will pay the price of a balanced budget. Equally problematic, Cardoso must do this while not contributing to the already massive social inequities in Brazil. He must face a powerful and unpleasant dilemma: how to cut government spending and improve the economy while not abandoning the economically disadvantaged.

Brazil has been a society with immense social and economic inequalities for centuries, and for the past one hundred years it has been able to alleviate social tensions, to some extent, through impressive economic growth. Between the 1870s and the 1980s, per-capita income in Brazil grew by 1,100 percent, a rate higher than any country in the world except Japan. Yet, in the 1980s, for the first time in more than a century, the economy and per-capita income stagnated. Continued stagnation could have profound repercussions in a society that has the most unequal income distribution in the world.

As the end of the twentieth century approaches, the flawed process of Brazilian industrialization has been all too apparent. Despite the enormous success over the last century in generating economic growth and transforming the nation from a coffee exporter to an emerging industrial power, the benefits of this industrialization have been distributed in an extremely uneven fashion both socially and regionally. More than 90 percent of the nation's industry is concentrated in the South and Southeast. While the richest 20 percent of the population earn 65 percent of the national income, the poorest 20 percent now earn less than 3 percent.

The enormous task that faces Brazil's political leadership is to restart the process of economic growth while dismantling those structures and institutions that so severely restrict the economic opportunities of the masses. Fernando Henrique Cardoso and Fernando Collor (despite his enormous flaws) seem to have definitively moved Brazil back onto the path to economic growth. The old nationalism and protectionism have been profoundly challenged, if not defeated. The state interventionism of the postwar era has outlived its usefulness and has now been attacked and discredited. As Collor and Cardoso discovered, dismantling the bloated bureaucratic machinery will be tough. As Eastern Europeans and citizens of the old Soviet Union are also discovering, both the state and the people find it hard to relinquish the old structures of intervention.

Like the people in the former Soviet bloc, the Brazilians confront a difficult and important transition. While the problems and failures of past practices may seem reasonably clear, the path to the future remains clouded and uncertain. In both places, the key to the future hinges on wise choices by political leaders. Both regions are struggling with the consequences of decades of political repression, although the problem is more severe in Eastern Europe. Brazil sorely needs a new generation of political leadership, a generation whose formation and experience has not been severely hampered by two decades of military rule.

Despite the gloomy picture of the last decade, I believe that the Brazilians will find a way to a prosperous and more equitable future. In spite of Brazil's huge problems, only a handful of countries are as blessed with resources or have a stronger industrial infrastructure. By no stretch of the imagination is the outlook for Brazil as depressing as that of countries like Ethiopia, El Salvador, or Cambodia. After a decade of the worst economic crisis in its history, Brazil still has the largest stock of foreign investment ($32 billion) of any country in the developing world. Furthermore, Brazil has managed to dig itself out of previous crises and resume economic growth at an even faster pace than before. In the mid-1960s, for example, when it appeared that Brazil was disintegrating, no one could have imagined the phenomenal economic growth that doubled the size of the Brazilian GDP between 1968 and 1974. Finally, Brazil also has a very large and sophisticated community of intellectuals, politicians, and social scientists. I believe that their intelligence, combined with the ingenuity and cunning (*jeito*) so characteristic of Brazilians, will enable the country to overcome this crisis.

The challenges are formidable, but not insurmountable. To succeed the Brazilians will need to elect political leaders who will look beyond the divisive interest-group politics of the past. Both the politicians and the people must be willing to make the sacrifices that the country needs to move away from a system that benefits the minority. It will be no easy task persuading the elites and the middle classes to bear the brunt of these sacrifices. In the long run, these sacrifices will benefit both the masses and the privileged minority. The failure to fundamentally redirect growth could mean a descent into chaos, which would leave Brazil's industrial economy strangled by a Third World social structure well into the next century. Success, however, would put Brazilians back on the path to development and give it the chance to confront the immense social problems of this democratic, industrial giant. Should they succeed, they could begin to fulfill the dream of moving Brazil into the ranks of the First World in the twenty-first century. For all of the industrial revolution's flaws, it can still succeed.

Once and Future

> What exists today is the concrete possibility that
> the future could become the big victim of the
> present.
> —José Serra

BRAZILIANS HAVE CREATED A RICH AND COMPLEX NATION that could
become a major player on the world stage in the twenty-first century.
With enormous natural and human resources, a strong industrial base,
and a dynamic culture, Brazil could play a prominent role in international politics, economic relations, and cultural affairs. It could become
the first developing nation to enter into the ranks of the great powers.
Brazil could leave the Third World.

The creation of Brazil has been no small accomplishment. Out of
the complex and complicated collision of Africa, Europe, and the
Americas, the Brazilians have forged a nation of substantial achievements and enormous potential. The creation of Brazil looks even more
impressive when compared with the struggles of other peoples to forge
their own societies and cultures. Few nations of continental dimensions
have been so successful in creating such a strong sense of national
identity.

Despite the persistent problems of racism, class inequity, and
political elitism, Brazil's problems often pale in comparison with other
large nations, especially those in the developed world. Racial mixture
over centuries has created an ethnic melting pot that frees Brazilians
from the bitter racial divisions found in so many societies. Brazil does
not have to contend with the racial divisions so characteristic of the
United States or South Africa. A common linguistic and religious

heritage brings Brazilians together, and offers a stark contrast to the fratricidal violence of a nation like India, which is divided by language, religion, and caste.

Although pronounced regional, social, economic, and political inequalities divide Brazilians, a single language, racial mixture, and religion–a common set of cultural traditions–brings them together as a nation. The Brazilians have truly created a national culture and a national identity. Out of the contentious clash of three civilizations has emerged a new civilization. In this sense, Gilberto Freyre had it right. Brazilians have fashioned a lusotropical civilization that is unique and that should be a source of great pride.

Although I have frequently compared Brazil to the United States throughout this book, my goal has not been to show how it has failed to be like the United States. I have tried to stress that Brazil has its own historical path, its own unique culture. Brazil is not simply a distorted or incomplete version of the United States. The two nations have traveled along similar paths that often parallel one another, intersect, and diverge. My efforts to compare them is grounded in the belief that peoples understand each other better through comparison rather than through isolated study. By looking for points of convergence and divergence, we come to know a good deal about Brazil, and ourselves, and we are richer for it.

Although I have refrained from writing a history text, the historical bias of this interpretive survey has also been conscious and deliberate. I have tried to explain contemporary Brazil, while always underscoring the need to recognize the dynamics of change. No large-scale society remains static. As the old maxim goes, the only constant in history is change. I have tried to emphasize both the continuities *and* the changes in Brazilian society and culture. Continuities define nations and peoples. They are the patterns and structures that we call culture. Yet, I believe that we should not be deceived by the continuities. Cultures are always changing. Brazil has changed dramatically over the past five hundred years.

Some prominent experts, Brazilian and foreign, have concluded that Brazil has changed more in form than in substance over the last five hundred years. The continuities have certainly been enduring and pronounced. The division between a small, wealthy elite and a huge, impoverished lower class has endured for centuries, despite changes in political regimes and despite economic modernization. Elite manipulation and domination of the political system has persisted although the power and cohesion of the elite have been severely challenged and altered during this century. The corporatist ethos continues to pervade much of Brazilian society and culture. Having stressed the continuities, however, I find the argument that "the more things change, the more they stay the same" to be simplistic and ahistorical.

Contemporary Brazil is a modern nation-state that cannot compare with the fragmented, sparsely populated, and loosely administered colonial enterprise that the Portuguese claimed to control up to the nineteenth century. Even in the late nineteenth century, Brazil remained an incipient nation-state. The creation of a national culture also did not come to fruition until this century. Finally, while elite power in politics, society, and the economy has persisted, the nature of that power has changed dramatically. The absolutism of the monarchy or the power of the imperial elite cannot compare with the efforts of a very diverse group of elites to assert their power in a democratic regime with mass media and more than 90 million voters. Elite power in a diverse and highly industrialized economy differs profoundly from the control large landowners exerted in an overwhelmingly agrarian and monocultural economy prior to the twentieth century. Brazil, in short, has been constantly evolving and changing in substantial and important ways.

The Brazilians have overcome great obstacles to construct their nation. Colonialism both created and handicapped Brazil. Without Portuguese imperial expansion and the African slave trade, there would be no Brazil; yet, colonialism saddled the new nation with a devastating legacy of racism, monoculture, class divisions, and political elitism. Over the last two centuries, the Brazilians have used the resources of the land and their own resourcefulness to whittle away at the colonial heritage. The legacy of colonialism hindered development and delayed Brazil's efforts to modernize. When the Brazilians began to make efforts to shed the colonial economic heritage, they faced a difficult struggle. The efforts of the North Atlantic nations to maintain economic superiority, and to control the Western economy, have made Brazil's struggle to industrialize and develop long and difficult.

While writing this book over the past eight years, I often wondered how I would be able to end it on an optimistic note. The crisis of the last decade frequently seemed to be a long nightmare from which Brazilians would never awaken. It seemed to have once again, perhaps permanently, derailed Brazil's emergence as one of the world's great nations. Brazil has been a nation poised on the brink of greatness for much too long. Many felt that the economic "miracle" of the 1970s would move Brazil close to the brink. In economic terms, Brazil had stepped beyond the ranks of the developing world by the 1980s, but it had not yet entered into the ranks of the developed world. After a decade of the most prolonged and severe economic crisis in the nation's history, pessimism pervaded the outlook of most observers. A survey of the most influential elite figures in Brazil in the late eighties showed that even the powerful did not see much hope for a breakthrough in the near future.

In the short term, it was often hard not to share this pessimism. In the 1980s and early 1990s, civilian political leaders failed to provide the country with direction and vision. Civilian political leaders failed miserably, and repeatedly, in their efforts to resolve the economic crisis, and few were willing to address the most basic of all issues—the social question. Brazil, alas, remains poised on the verge of greatness. It remains the country of the future.

As a historian taking the "longer view," I remain optimistic about Brazil in the long term. We must not be overwhelmed by recent events. Brazil has faced severe crises before, and gone on to greater achievements. On the eve of the military coup in 1964, for example, when Brazil experienced a disintegrating economy and political chaos, few could have foreseen the extraordinary economic surge just a few years away or the return to democratic politics. The past fifteen years have been dark times for Brazilians, but I believe they will survive them and thrive. I hope and wish that the election of Fernando Henrique Cardoso and the changes that are taking place are fundamental and will be for the better. Economic growth and democratic politics are reemerging stronger than at any other time in Brazilian history. The pursuit of social equity will be tougher, and it will be the true measure of Brazil's entry into the ranks of the developed nations.

The fate of Brazil, more so today than at any other time in its history, is in the hands of Brazilians. The great powers still dominate and shape the international economic and political arena, but Brazil has significantly altered its position in the world during the past half-century. The Brazilians are neither autonomous nor free to dictate their own future, but they now have a greater ability to shape it than ever before.

Brazilians must draw on their best qualities to realize their potential. They must compromise and conciliate to find creative ways to begin to break down the profound social apartheid and elitism that plague the nation. Should they fail to move toward more democratic politics and greater social equity, Brazil could turn into one of the greatest failures of the modern era. Brazil could enter the twenty-first century with a First World industrial economy strangled by a Third World social structure.

Bolstered by the seemingly irrepressible optimism of Brazilians, I believe that they will find ways to achieve sustained economic growth and to create the wealth for development. They will work out political solutions to their profound social problems and make development a reality. I hope, and I believe, that Brazil will finally achieve its potential and no longer be condemned to be the country of the future.

APPENDIX A

A Statistical Profile of Brazil's Regions

	NE	SE	S	N	CW	Brazil
Area (% of total area)	18.27	10.85	6.76	45.26	18.86	100.00
Population (% of total population, 1991)	28.94	42.73	15.08	6.83	6.42	100.00
Life Expectancy (1990)	64.22	65.53	68.68	67.5	67.8	65.62
Infant Mortality (per 1,000 live births, 1990)	88.2	30.0	26.7	53.2	33.0	49.7
Literacy (% of persons over age 7, 1990)	62.5	88.5	90.2	75.1	84.4	80.3
Poverty Index (1990, % of economically active population making less than twice minimum wage)	84.7	63.4	67.6	68.3	68.3	70.5

Source: Fundação Instituto Brasileiro de Geografia e Estatística, *Anuário estatístico do Brasil, 1995* (Rio de Janeiro: IBGE, 1996).

NE = Northeast: Maranhão, Piauí, Ceará, Rio Grande do Norte, Paraíba, Pernambuco, Alagoas, Sergipe, Bahia
SE = Southeast: Minas Gerais, Espírito Santo, Rio de Janeiro, São Paulo
S = South: Paraná, Santa Catarina, Rio Grande do Sul
N = North: Rondônia, Acre, Amazonas, Roraima, Pará, Amapá, Tocantins
CW = Center-West: Mato Grosso do Sul, Mato Grosso, Goiás, Distrito Federal

Note: In 1990, twice the minimum wage equalled roughly $100 per month.

APPENDIX B

Population of Brazil

Year	Population
1776	1,900,000
1808	2,419,406
1819	3,596,132
1823	3,960,860
1830	5,340,000
1854	7,677,800
1872	9,930,478
1890	14,333,915
1900	17,438,434
1920	30,635,605
1930	37,625,436
1940	41,236,315
1950	51,944,397
1960	70,191,370
1970	93,139,037
1980	119,098,992
1985	135,564,400
1991	146,825,475
1995	155,822,400 (est.)

Sources: Armin K. Ludwig, Brazil: *A Handbook of Historical Statistics* (Boston: G. K. Hall, 1985), p. 47. Fundação Instituto Brasileiro de Geografia e Estatística, *Brasil em números,* v. 4, 1995/1996 (Rio de Janeiro: IBGE, 1995/1996).

Note: Census figures, especially prior to 1872, are notoriously unreliable, and generally are based on projections and estimates.

APPENDIX C

Population of Major Brazilian Cities, 1991

City	Municipality	Metropolitan Region
São Paulo	9,646,185	15,416,410
Rio de Janeiro	5,480,768	9,796,498
Belo Horizonte	2,020,161	3,431,756
Porto Alegre	1,263,403	3,026,029
Recife	1,298,229	2,871,261
Salvador	2,075,273	2,493,224
Fortaleza	1,768,637	2,303,645
Curitiba	1,315,035	1,998,807
Brasília	1,601,094	—
Belém	1,244,689	1,332,723
Manaus	1,011,501	—

Sources: *Anuário estatístico do Brasil* (Rio de Janeiro: IBGE, 1995), p. 2-14; *Brasil, favelas, 1980 e 1991* (Rio de Janeiro: IBGE, 1995).

APPENDIX D

Income Distribution in Brazil

Income Groups (Percentile: Lowest Income Group to Highest)	Relative Income Levels (Percent of Total Income)			
	1960	1970	1980	1995
1-10	1.9	1.2	1.1	1.1
11-20	2.0	2.0	2.0	2.2
21-30	3.0	3.0	3.0	2.4
31-40	4.4	3.8	3.5	3.2
41-50	6.1	5.0	4.5	4.2
51-60	7.5	6.2	5.5	5.3
61-70	9.0	7.2	7.2	7.2
71-80	11.3	10.0	9.6	10.1
81-90	15.2	15.1	15.3	16.1
91-100	39.6	46.5	48.3	48.2
	100.0	100.0	100.0	100.0
Bottom 50 percent	17.4	15.0	14.1	13.1
Top 5 percent	28.3	34.1	37.9	34.6
Top 1 percent	11.9	14.7	16.9	13.9

Sources: James Lang, *Inside Development in Latin America* (Chapel Hill, NC: University of North Carolina Press, 1988), p. 151; *Pesquisa Nacional por Amostra de Domicílios—PNAD: síntese de indicadores 1995* (Rio de Janeiro: IBGE, 1996), p. 73.

NOTES

In writing this book for the non-specialist I have consciously avoided the use of lots of footnotes and standard scholarly citations. The following are the sources just for the direct quotes in the text.

vi, *These people* . . . : Roberto DaMatta, *Carnivals, Rogues, and Heroes: An Interpretation of the Brazilian Dilemma,* trans. John Drury (South Bend, IN: University of Notre Dame Press, 1991), 3.

PREFACE

xi, *Brazil is not* . . . : quoted in Roberto DaMatta, *Conta de mentiroso: sete ensaios de antropologia brasileira* (Rio de Janeiro: Rocco, 1993), 7.

CHAPTER 1. THE PRESENCE OF THE PAST

7, *I live in the* . . . : from the poem, "Que país é este?" as cited in E. Bradford Burns, *A History of Brazil,* 3rd ed. (New York: Columbia University Press, 1993), 475.

10, *in the service of God* . . . : quoted in Benjamin Keen, *A History of Latin America,* 5th ed. (Boston: Houghton Mifflin, 1996), vol. I, 77.

12, *the greatest event* . . . : quoted in Keen, *A History of Latin America,* vol. I, 59.

15, *naked and without* . . . : Pedro Vaz de Caminha quoted in John Hemming, *Red Gold: The Conquest of the Brazilian Indians, 1500-1760* (Cambridge, MA: Harvard University Press, 1978), 3.

17, *to draw from* . . . : quoted in Hemming, *Red Gold,* epigraph facing page 1.

20, *cling crablike to the* . . . : quoted in Burns, *A History of Brazil,* 10.

22, *They go without* . . . : Hemming, *Red Gold,* 246.

31, *He knew better* . . . : quoted in C. H. Haring, *Empire in Brazil: A New World Experiment with Monarchy* (Cambridge, MA: Harvard University Press, 1958), 42.

35, *Love as the* . . . : Auguste Comte, *Catéchisme positiviste* (Paris, 1852), title page.

49, *that shot was aimed* . . . : quoted in Thomas E. Skidmore, *Politics in Brazil, 1930-1964: An Experiment in Democracy* (London: Oxford University Press, 1967), 138.

49, *I gave you my life.* . . . : quoted in E. Bradford Burns, *A Documentary History of Brazil* (New York: Alfred A. Knopf, 1966), 370-71.

51, *a politician who* . . . : Skidmore, *Politics in Brazil,* 188.

65, *My past is not* . . . : from Carlos Guilherme Mota and Adriana Lopez, eds., *Brasil revisitado: palavras e imagens* (São Paulo: Editora Rios, 1889), 151.

CHAPTER 2. THE BRAZILIAN ARCHIPELAGO

67, *The astonishing* . . . : from *The Gilberto Freyre Reader,* trans. Barbara Shelby (New York: Alfred A. Knopf, 1974), 43.

79, *two blocks of Paris* . . . : Denny Braun, *The Rich Get Richer* (Chicago: Nelson-Hall, 1991), 90.

80, *natural beauty* . . . : Roger Bastide, *Brasil: terra de constrastes,* 5th ed., trans. Maria Isaura Pereira Queiroz (São Paulo: Difusão Européia do Livro, 1973), 144.

85, *northeastern coast.* Bastide, *Brasil,* 173.

87, *last unwritten page* . . . : quoted in Susanna Hecht and Alexander Cockburn, *The Fate of the Forest: Developers, Destroyers and Defenders of the Amazon* (New York: Harper Perennial, 1990), 16.

89, *unsolved mysteries* . . . : Edward O. Wilson, *The Diversity of Life* (Cambridge, MA: Belknap Press of Harvard University Press, 1992), 7.

89, *We ourselves* . . . : from José Toribio Medina, ed., *The Discovery of the Amazon* (New York: Dover, 1988), 214.

95, *Die if necessary...* : quoted in Alex Shoumatoff, *The Rivers Amazon* (San Francisco: Sierra Club Books, 1986), 39.

96, *victims of the ...* : Shelton H. Davis, *Victims of the Miracle: Development and the Indians of Brazil* (Cambridge: Cambridge University Press, 1977).

100, high-tech... : Shoumatoff, Rivers Amazon, xi.

CHAPTER 3. LUSOTROPICAL CIVILIZATION

103, *The formation...* : Gilberto Freyre, *The Masters and the Slaves: A Study in the Development of Brazilian Civilization,* 2nd ed. rev., trans. Samuel Putnam (New York: Alfred A. Knopf, 1970), 79.

117, *new world in the...* : Gilberto Freyre, *New Work in the Tropics: The Culture of Modern Brazil* (New York: Vintage, 1963).

124, *The anthropologist Roberto DaMatta ...* : Roberto DaMatta, *Carnivals, Rogues, and Heroes: An Interpretation of the Brazilian Dilemma,* trans. John Drury (South Bend, IN: University of Notre Dame Press, 1991), chapter 4, 137-197.

126, *white mask over a...* : Bastide, *Brasil: terra de constrastes,* 5th ed., trans. Maria Isaura Pereira Queiroz (São Paulo: Difusão Européia do Livro, 1973), 74.

126, *is a genuine bit of...* : Roger Bastide, *The African Religions of Brazil: Toward a Sociology of the Interpenetration of Civilizations,* trans. Helen Sebba (Baltimore, MD: The Johns Hopkins University Press, 1978), 224.

128, *The Rio* macumba... : Bastide, *The African Religions of Brazil,* 298.

129, *it is possible in Brazil...* : Bastide, *The African Religions of Brazil,* 28.

136, *Mulatas are glorified ...* : Alma Guillermoprieto, *Samba* (New York: Alfred A. Knopf, 1990), 180.

140, *As long as you are...* : quoted in Paul Rambali, *In the Cities and Jungles of Brazil* (New York: Henry Holt and Company, 1993), 57.

143, *the hard reality ...* : Roberto DaMatta, *Carnaval as a Cultural Problem: Towards a Theory of Formal Events and Their Magic* (South Bend, IN: Helen Kellogg Institute for International Studies, 1986), 13.

143, *web of obligatory*...: DaMatta, *Carnaval as a Cultural Problem*, 11.

143, *Encouraging people*...: Ian Buruma, *A Japanese Mirror: Heroes and Villains of Japanese Culture* (London: Jonathan Cape), 224.

144, *is the glorification*...: DaMatta, *Carnivals, Rogues, and Heroes*, 86.

147, *which has the richest*...: *La Razón*, Buenos Aires, 11 April 1982.

157, *the greatest author*...: Susan Sontag, "Afterlives: The Case of Machado de Assis," *New Yorker* (7 May 1990), 107.

158, *We are condemned*...: Euclides da Cunha, *Rebellion in the Backlands (Os Sertões)*, trans. Samuel Putnam (Chicago: University of Chicago Press, 1944), 54.

158, *We are the sons*...: quoted in Burns, *A History of Brazil*, 3rd ed. (New York: Columbia University Press, 1993), 327-28.

159, *All history*...: from the poem, "Museu da Inconfidência," Carlos Drummond de Andrade, *Poesia completa e prosa* (Rio de Janeiro: Companhia José Aguilar Editora, 1973), 257.

CHAPTER 4. POWER AND PATRONAGE

165, *Brazil is a nation*...: in Kalman H. Silvert, *Expectant Peoples: Nationalism and Development* (New York: Random House, 1963), 232.

168, *man makes and is*...: Richard M. Morse, *New World Soundings: Culture and Ideology in the Americas* (Baltimore, MD: Johns Hopkins University Press, 1989), 127.

170, *institutionalize an individualistic*...: DaMatta, *Carnivals, Rogues, and Heroes: An Interpretation of the Brazilian Dilemma*, trans. John Drury (South Bend, IN: University of Notre Dame Press, 1991), xi.

175, *go to mass*: Gabriel García Márquez, *One Hundred Years of Solitude*, trans. Gregory Rabassa (New York: Avon, 1971), 228.

184, *delegated democracy*...: Guillermo A. O'Donnell, "Democracia delegativa?" *Cuadernos del CLAEH*, 61 (July 1992), 5-20.

189, *middle-class military*: José Nun, "A Latin American Phenomenon: The Middle-Class Military Coup," in James Petras and Maurice Zeitlin, eds., *Latin America: Reform or Revolution? A Reader* (Greenwich, CT: Fawcett Publications, 1968), 145-85.

193, *The Spirit of the Lord* . . . : Luke 4:18, *The Holy Bible, King James Version* (Cleveland, OH: World Publishing Company, n.d.).

205, *We made a fire* . . . : quoted in Kurt Weyland, "The Rise and Fall of President Collor and its Impact on Brazilian Democracy," *Journal of Interamerican Studies and World Affairs*, 35:1 (1993), 19.

208, *illuminate the moral* . . . : quoted in Alan Cowell, "Pope's Tight Limits," *New York Times* (16 October 1991), A6.

CHAPTER 5. A FLAWED INDUSTRIAL REVOLUTION

211, *Some people think* . . . : quoted in DaMatta, *Carnivals, Rogues, and Heroes: An Interpretation of the Brazilian Dilemma*, trans. John Drury (South Bend, IN: University of Notre Dame Press, 1991), 3.

212, *Belindia*: Edmar Lisboa Bacha, "El economista y el rey de Belindia: una fábula para tecnocratas," *El Trimestre Económico*, 42:167 (July-September 1975), 725-30.

245, *integrar para não* . . . : Ariovaldo Umbelino de Oliveira, *Integrar para não entregar: políticas públicas e Amazônia* (Campinas, SP: Papirus, 1988).

247, *Until very recently* . . . : João Paulo Velloso quoted in *Time* (22 May 1972), 45.

CONCLUSION

259, *What exists today* . . . : José Serra, "Existe uma saída," *Veja* (1 August 1990), 58.

SUGGESTED READINGS

The following list contains books in English that I believe will be useful and informative for the non-specialist. I have tried to be very selective, leaving out a great number of books and articles that proved useful in writing this book because they were written primarily for academic specialists. The bibliography—like this book—is multidisciplinary. For those interested in a more detailed and more fully annotated bibliography, I suggest that you write to the Latin American Institute at the University of New Mexico in Albuquerque. Over the last few years, the Institute has published a series of booklets containing bibliographical essays on all aspects of Brazilian studies.

1. THE PRESENCE OF THE PAST

Over the past thirty years, a large number of works have appeared in English on Brazilian history. E. Bradford Burns, *A History of Brazil*, remains the best general text available, and his *A Documentary History of Brazil* contains an excellent collection of primary sources translated into English. Ronald M. Schneider's recent *"Order and Progress"* takes the very traditional approach of history as politics and does not have the wide-ranging social and cultural perspective found in Burns. Those interested in intellectual history should consult João Cruz Costa, *A History of Ideas in Brazil*.

The best general overview of colonial Brazil is James Lang, *Portuguese Brazil*. Leslie Bethell has edited two important volumes, *Colonial Brazil* and *Brazil: Empire and Republic, 1822-1930* that contain the latest scholarly research and excellent bibliographical essays. C. R. Boxer, a British historian, has written a number of the most important works on colonial Brazil. *The Portuguese Seaborne Empire, 1415-1825* offers a wonderful introduction to Portugal's role in European overseas expansion. *The Dutch in Brazil, 1624-1654* and *The Golden Age of*

Brazil, 1695-1750 are two indispensable works on seventeenth- and eighteenth-century Brazil. *The Bandeirantes* contains a collection of accounts by pioneers into the Brazilian interior during the colonial period with a fine introductory essay by Richard Morse.

Alfred Crosby's *The Columbian Exchange* examines the larger biological and cultural exchange between the Old World and the New, providing a fascinating look at the impact on both worlds. John Hemming's *Red Gold* chronicles the devastating consequences of European expansion on Brazil's Indian peoples through the late eighteenth century. A second volume, *Amazon Frontier,* takes the story up to the early twentieth century. The best account of the European encounters with Brazilian Indians in the sixteenth century is Jean de Léry's *History of a Voyage to the Land of Brazil,* originally published in 1578, and recently translated into English.

Although dated, Caio Prado Júnior, *The Colonial Background of Modern Brazil,* remains a very useful introduction to the period. Stanley and Barbara Stein, *The Colonial Heritage of Latin America,* is a provocative and challenging overview of the colonial period that places Brazil within the context of Latin American history. Gilberto Freyre's landmark *The Masters and the Slaves* offers a brilliant, yet flawed, interpretation of the construction of Brazilian society. Full of insights and provocative assertions, this book has shaped Brazilians' perception of their own history more than any other work in this century. *New World in the Tropics* offers a more concise introduction to Freyre's views.

The best general overview of slavery is Katia Mattoso, *To Be a Slave in Brazil.* Robert Conrad has edited an exceptional collection of documents on slavery and the slave trade with the marvelous title *Children of God's Fire.* Carl Degler's *Neither Black Nor White* is an indispensable comparison of slavery and race relations in the United States and Brazil. Robert Toplin, *The Abolition of Slavery in Brazil,* is the best general work on the subject.

A number of important works on imperial Brazil have appeared recently. Roderick Barman's *Brazil: The Forging of a Nation, 1798-1852* and Neill Macaulay's *Dom Pedro* are the best works on the first half of the century. Emilia Viotti da Costa, *The Brazilian Empire,* contains some penetrating scholarly essays on various aspects of the empire. A more traditional survey of the empire that remains useful and readable despite its age is C. H. Haring, *Empire in Brazil.* Richard Graham's *Britain and the Onset of Modernization in Brazil, 1850-1914* looks at the profound impact of Britain on nineteenth-century Brazil. Stanley Stein's *Vassouras* is a fine study of how coffee's expansion and decline transformed one county in the state of Rio de Janeiro during the nineteenth century.

For those interested in the inner workings of regionalism and politics during the First Republic, there are a series of studies on Brazilian states during this period by Robert M. Levine (Pernambuco), Joseph L. Love (Rio Grande do Sul and São Paulo), John D. Wirth (Minas Gerais), and Eul-Soo Pang (Bahia). Warren Dean, *Brazil and the Struggle for Rubber,* is a fine account of the Amazon rubber boom and bust. *The Prestes Column* by Neill Macaulay is a good introduction to the changes taking place in Brazil in the 1920s and the *tenentes* movement.

Two works by Thomas Skidmore, *Politics in Brazil, 1930-1964* and *The Politics of Military Rule in Brazil, 1964-85,* offer the best synthesis of Brazilian history since 1930, with a heavy emphasis on politics and economic policymaking. John W. F. Dulles has written a detailed, but not very exciting, biography of Getúlio Vargas, who still awaits his definitive biographer. A challenging and provocative interpretation of recent Brazilian history by a prominent economist and former cabinet minister is Luiz Carlos Bresser Pereira, *Development and Crisis in Brazil, 1930-1983.*

2. THE BRAZILIAN ARCHIPELAGO

There are no general surveys in English of Brazil's regions. I have learned a great deal from the insights of the French sociologist Roger Bastide. His *Brésil: Terre des Contrastes* has never been translated into English (although there is a Portuguese-language edition). Like a number of Frenchmen in the 1930s and 1940s, Bastide came under the spell of Brazil, especially its lush, sensuous, African side. His works contain some of the most insightful writing ever done by a foreigner about Brazil. His compatriot, the anthropologist Claude Lévi-Strauss, also first arrived in Brazil in the 1930s. His *Tristes Tropiques* is a fascinating account of encounters with Indians in the interior, and one of the great travel books of the twentieth century.

The travel literature on Brazil is substantial, and for the Amazon it is vast. I will single out just a few notable works. Alex Shoumatoff, *The*

Capital of Hope, looks at Brasília and its people, and *The Rivers Amazon* offers a perceptive portrayal of the region. Two of the best recent works on the Amazon are Susanna Hecht and Alexander Cockburn, *The Fate of the Forest*, and Adrian Cowell, *The Decade of Destruction*. Cowell is a fine British filmmaker who has been working in the Amazon for four decades.

3. LUSOTROPICAL CIVILIZATION

The bibliography on Brazilian society and culture is vast and varied. Although dated, the best general introduction to Brazilian society is Charles Wagley, *An Introduction to Brazil*. An anthropologist who married a Brazilian and spent much of his adult life in Brazil, Wagley had a formidable understanding of Brazilian culture. The many works of Gilberto Freyre remain important and stimulating introductions to Brazilian civilization, but they should be used with caution. His optimistic view of race relations, and his rather static view of Brazilian society, have been severely critiqued. José Honório Rodrigues, *The Brazilians*, offers the views of a prominent Brazilian historian on his society. A very useful and interesting comparison of social behavior in the United States and Brazil is Phyllis A. Harrison, *Behaving Brazilian*. Priscilla Goslin's *How to Be a Carioca* is a wonderfully humorous look at social customs in Rio de Janeiro (and Brazil).

Carolina Maria de Jesus, *Child of the Dark*, is a graphic and compelling firsthand account of poverty in Brazil. A black single mother of three small children, with a second-grade education, De Jesus kept a diary of her daily struggle to survive in the slums of São Paulo in the 1950s. Originally published in 1960, this diary became the all-time bestselling book in Brazil in the early sixties. This is a view of industrialization and modernization from the perspective of the poor. James Lang's *Inside Development in Latin America* contains sensitive portrayals of grassroots efforts to provide better health care for the poor in Brazil (as well as in Colombia and the Dominican Republic).

Degler's *Neither Black Nor White* is the best survey of race relations in English. Florestan Fernandes, *The Negro in Brazilian Society*, is an important survey by a very prominent Brazilian sociologist. For a survey of elite attitudes about race at the turn of the century, see Thomas E.

Skidmore, *Black into White.* The epic account of the Canudos rebellion by Euclídes da Cunha, *Rebellion in the Backlands,* is a monument in Brazilian letters and gives the best insight into elite attitudes on race and national identity at the turn of the century. David T. Haberly looks at race and identity from a literary perspective in *Three Sad Races.*

Roberto DaMatta, a Brazilian anthropologist, is one of the most astute observers of Brazilian life. I highly recommend his collection of essays, *Carnivals, Rogues, and Heroes.* I am also deeply indebted to the work of Richard Morse, who knows Brazil, perhaps, better than any other Brazilianist in this country. Some of Morse's essays are collected in *New World Soundings.* Richard Parker's *Bodies, Pleasures, and Passions* analyzes contemporary Brazilian sexual culture from the perspective of an anthropologist living in Brazil.

Bastide's *The African Religions of Brazil* is the best introduction to the topic. Seth and Ruth Leacock's *Spirits of the Deep* is an in-depth study of one Afro-Brazilian cult in northern Brazil. David Martin, *Tongues of Fire,* documents the explosive rise of Protestantism in Brazil and the rest of Latin America during the past few decades.

Alma Guillermoprieto's *Samba* presents a fascinating account of one samba school and its preparation for carnival. The book also offers insights into some of the nuances of Brazilian culture, especially race and sexual relations. Her essays on Brazil in *The Heart That Bleeds* are some of the most astute and observant writing on contemporary Brazil.

The best introduction to the Brazilian music scene is Chris McGowan and Ricardo Pessanha's *The Brazilian Sound.* Two good recent surveys of Brazilian music are David P. Appleby, *The Music of Brazil,* and Charles A. Perrone, *Masters of Contemporary Brazilian Song.* Randal Johnson has published a series of fine works on Brazilian cinema, including *Cinema Novo x 5* and *The Film Industry in Brazil.*

I have been hard pressed to condense Brazilian literature into a few pages. The bibliographical guide to Brazilian literature, put out by the Latin American Center at the University of New Mexico, is a good starting point for anyone seeking a brief bibliographical survey of the literature. I have chosen to single out just a few of the many great works in Brazilian literature. Machado de Assis' *Epitaph of a Small Winner,* or any of his other novels or short stories, is as a fine starting point as any. All of the many novels of Jorge Amado have been translated into English. I highly recommend three: *Gabriela, Clove and Cinnamon,* a lush and exuberant look at life in Bahia; *Tent of Miracles,* a wonderful and witty piece of historical fiction on race relations in the Northeast; and *The Violent Land,* an epic story of the expansion of cacao plantations in the early twentieth century. Graciliano Ramos, *Barren Lives,* is a spare and moving account of life in the drought-stricken interior of the Northeast. Márcio Souza has written a witty, satirical account of the rubber boom and the struggle over the territory of Acre in *Emperor of the Amazon.* For the more adventurous, there is João Guimarães Rosa's *Devil to Pay in the Backlands,* a complex and rich novel. The American poet Elizabeth Bishop edited *An Anthology of Twentieth-Century Brazilian Poetry,* and *Traveling in the Family* offers a selection of poems by Carlos Drummond de Andrade. John Nist, *The Modernist Movement in Brazil,* is a good introduction to this important cultural movement.

The best surveys in English of Brazil's love affair with soccer are the sociological study of Janet Lever, *Soccer Madness* and Tony Mason's *Passion of the People? Football in South America.*

4. POWER AND PATRONAGE

Although I think it is sometimes simplistic and ahistorical, the best introduction to Brazilian politics is Riordan Roett, *Brazil: Politics in a Patrimonial Society.* Peter Flynn combines a comprehensive political history and analysis in *Brazil: A Political Analysis.* The military regime and redemocratization have generated a large literature. The Skidmore volumes (noted above for chapter 1) are invaluable narratives of events, while the works of Maria Helena Moreira Alves and Alfred Stepan are good introductions to the military regime and the return to civilian politics. Márcio Moreira Alves (Maria Helena's father) has written a moving account of his opposition to the regime and his life underground in *A Grain of Mustard Seed.* The refusal of the congress to expel Alves set the stage for the military coup of the hardliners in December 1968.

Paulo Freire's widely influential *Pedagogy of the Oppressed* describes the methods and philosophy of this leader in the movement to teach the poor to read and write. Freire's methods have been widely copied around the world. A good treatment of the life and death of Chico Mendes and the destruction of the rain forest is Andrew Revkin's *The Burning Season.*

There are a number of good books on the Catholic Church in contemporary Brazilian politics. Penny Lernoux looks at the activist church and liberation theology in all of Latin America, with extensive coverage of Brazil, in *Cry of the People*. Scott Mainwaring, a political scientist, traces the changing role of the church in *The Catholic Church and Politics in Brazil, 1916-1985*. *Torture in Brazil* is the edited translation of *Nunca Mais*, the report on torture under military rule that was compiled by the archdiocese of São Paulo. Lawrence Weschler's *A Miracle, A Universe*, describes the fascinating process of putting the report together, and he compares the ways in which Uruguayans and Brazilians attempt to settle accounts with the torturers.

Robert Wesson, *The United States and Brazil*, and Jan Knippers Black, *United States Penetration of Brazil*, analyze U.S.-Brazilian relations from two very different perspectives.

5. A FLAWED INDUSTRIAL REVOLUTION

The best introduction to the Brazilian economy, and Brazilian economic history, is Werner Baer, *The Brazilian Economy*. Warren Dean traces the impressive rise of Brazil's biggest city in *The Industrialization of São Paulo, 1880-1945*. Peter Evans, *Dependent Development*, presents a challenging and sophisticated analysis of recent industrialization by a prominent sociologist. A good overview of the origins and nature of the debt crisis is Pedro-Pablo Kuczynski, *Latin American Debt*.

BIBLIOGRAPHY

Alves, Márcio Moreira. *A Grain of Mustard Seed: The Awakening of the Brazilian Revolution.* Garden City, NY: Anchor Books, 1973.

Alves, Maria Helena Moreira. *State and Opposition in Military Brazil.* Austin: University of Texas Press, 1985.

Amado, Jorge. *Gabriela, Clove and Cinnamon.* Trans. James L. Taylor and William L. Grossman. New York: Avon, 1962.

———. *Tent of Miracles.* Trans. Barbara Shelby. New York: Avon, 1971.

———. *The Violent Land.* Trans. Samuel Putnam. New York: Alfred A. Knopf, 1988.

Andrade, Carlos Drummond de. *Traveling in the Family: Selected Poems.* Edited by Thomas Colchie and Mark Strand. New York: Random House, 1986.

Appleby, David P. *The Music of Brazil.* Austin: University of Texas Press, 1983.

Baer, Werner. *The Brazilian Economy: Growth and Development,* 4th ed. Westport, CT: Praeger, 1995.

Barman, Roderick. J. *Brazil: The Forging of a Nation, 1798-1852.* Stanford: Stanford University Press, 1988.

Bastide, Roger. *The African Religions of Brazil: Toward a Sociology of the Interpenetration of Civilizations.* Trans. Helen Sabba. Baltimore: Johns Hopkins University Press, 1978.

Bethell, Leslie. ed. *Brazil: Empire and Republic, 1822-1930.* Cambridge: Cambridge University Press, 1989.

———. ed. *Colonial Brazil.* Cambridge: Cambridge University Press, 1987.

Bishop, Elizabeth and Emanuel Brasil. eds. *An Anthology of Twentieth-Century Brazilian Poetry.* Middletown, CT: Wesleyan University Press, 1972.

Black, Jan Knippers. *United States Penetration of Brazil.* Philadelphia: University of Pennsylvania Press, 1977.

Boxer, C. R. *The Dutch in Brazil, 1624-1654.* Oxford: Oxford University Press, 1957.

———. *The Golden Age of Brazil, 1695-1750: Growing Pains of a Colonial Society.* Berkeley: University of California Press, 1969.

———. *The Portuguese Seaborne Empire, 1415-1825.* New York: Alfred A. Knopf, 1969.

Burns, E. Bradford. ed. *A Documentary History of Brazil.* New York: Alfred A. Knopf, 1966.

Burns, E. Bradford. *A History of Brazil,* 3rd ed. New York: Columbia University Press, 1993.

Catholic Church. Archdiocese of São Paulo, Brazil. *Torture in Brazil.* Trans. Jaime Wright. Ed. Joan Dassin. New York: Vintage Books, 1986.

Conrad, Robert E. *Children of God's Fire: A Documentary History of Black Slavery in Brazil.* Princeton: Princeton University Press, 1983.

Costa, Emília Viotti da. *The Brazilian Empire: Myths and Histories.* Chicago: University of Chicago Press, 1985.

Costa, João Cruz. *A History of Ideas in Brazil.* Trans. Suzette Macedo. Berkeley: University of California Press, 1964.

Cowell, Adrian. *The Decade of Destruction: The Crusade to Save the Amazon Rain Forest.* New York: Henry Holt, 1990.

Crosby, Alfred W., Jr. *The Columbian Exchange: Biological and Cultural Consequences of 1492.* Westport, CT: Greenwood Press, 1972.

Cunha, Euclídes da. *Rebellion in the Backlands.* Trans. Samuel Putnam. Chicago: University of Chicago Press, 1944.

DaMatta, Roberto. *Carnivals, Rogues, and Heroes: An Interpretation of the Brazilian Dilemma.* Trans. John Drury. Notre Dame, IN: University of Notre Dame Press, 1991.

Dean, Warren. *Brazil and the Struggle for Rubber: A Study in Environmental History.* Cambridge: Cambridge University Press, 1987.

———. *The Industrialization of São Paulo, 1880-1945.* Austin: University of Texas Press, 1969.

Degler, Carl N. *Neither Black Nor White: Slavery and Race Relations in Brazil and the United States.* Madison, WI: University of Wisconsin Press, 1986.

De Jesus, Carolina Maria. *Child of the Dark.* Trans. David St. Clair. New York: Mentor, 1962.

Dulles, John W. F. *Vargas of Brazil: A Political Biography.* Austin: University of Texas Press, 1967.

Evans, Peter. *Dependent Development: The Alliance of Multinational, State and Local Capital in Brazil.* Princeton: Princeton University Press, 1979.

Fernandes, Florestan. *The Negro in Brazilian Society.* Trans. Jacqueline D. Skiles, A. Brunel, and Arthur Rothwell. Ed. Phyllis B. Eveleth. New York: Columbia University Press, 1969.

Flynn, Peter. *Brazil: A Political Analysis.* Boulder, CO: Westview Press, 1979.

Freire, Paulo. *Pedagogy of the Oppressed.* New York: Seabury, 1970.

Freyre, Gilberto. *The Masters and the Slaves: A Study in the Development of Brazilian Civilization.* Trans. Samuel Putnam. New York: Alfred A. Knopf, 1956.

———. *New World in the Tropics: The Culture of Modern Brazil.* New York: Alfred A. Knopf, 1959.

Goslin, Priscilla Ann. *How to Be a Carioca: An Alternative Guide for the Tourist in Rio.* Rio de Janeiro: Two Can Press, 1993.

Graham, Richard. *Britain and the Onset of Modernization in Brazil, 1850-1914.* London: Cambridge University Press, 1968.

Guillermoprieto, Alma. *Samba.* New York: Alfred A. Knopf, 1990.

Haberly, David T. *Three Sad Races: Racial Identity and National Consciousness in Brazilian Literature.* Cambridge: Cambridge University Press, 1983.

Haring, C. H. *Empire in Brazil: A New World Experiment with Monarchy.* Cambridge, MA: Harvard University Press, 1958.

Harrison, Phyllis A. *Behaving Brazilian: A Comparison of Brazilian and North American Social Behavior.* Cambridge, MA: Newbury House, 1983.

Hecht, Susanna and Alexander Cockburn. *The Fate of the Forest: Developers, Destroyers and Defenders of the Amazon.* New York: Harper, 1990.

Hemming, John. *Amazon Frontier: The Defeat of the Brazilian Indians.* Cambridge, MA: Harvard University Press, 1987.

———. *Red Gold: The Conquest of the Brazilian Indians, 1500-1760.* Cambridge, MA: Harvard University Press, 1978.

Johnson, Randal. *Cinema Nova x 5: Masters of Contemporary Brazilian Film.* Austin: University of Texas Press, 1984.

———. *The Film Industry in Brazil: Culture and the State.* Pittsburgh: University of Pittsburgh Press, 1987.

Kuczynski, Pedro-Pablo. *Latin American Debt.* Baltimore: Johns Hopkins University Press, 1988.

Lang, James. *Inside Development in Latin America: A Report from the Dominican Republic, Colombia, and Brazil.* Chapel Hill, NC: University of North Carolina Press, 1988.

———. *Portuguese Brazil: The King's Plantation.* New York: Academic Press, 1979.

Leacock, Seth and Ruth. *Spirits of the Deep: A Study of an Afro-Brazilian Cult.* Garden City, NY: Doubleday, 1972.

Lernoux, Penny. *Cry of the People: The Struggle for Human Rights in Latin America— The Catholic Church in Conflict with U.S. Policy.* New York: Penguin Books, 1982.

Léry, Jean de. *History of a Voyage to the Land of Brazil.* Trans. Janet Whatley. Berkeley: University of California Press, 1990.

Lever, Janet. *Soccer Madness.* Chicago: University of Chicago Press, 1983.

Levine, Robert M. *Pernambuco and the Brazilian Federation, 1889-1937.* Stanford: Stanford University Press, 1978.

Lévi-Strauss, Claude. *Tristes Tropiques.* Trans. John and Doreen Weightman. New York: Atheneum, 1974.

Love, Joseph L. *Rio Grande do Sul and Brazilian Regionalism: 1882-1930.* Stanford: Stanford University Press, 1971.

———. *São Paulo and the Brazilian Federation, 1889-1937.* Stanford: Stanford University Press, 1980.

Macaulay, Neill. *Dom Pedro: The Struggle for Liberty in Brazil and Portugal, 1798-1834.* Durham, NC: Duke University Press, 1986.

———. *The Prestes Column: Revolution in Brazil.* New York: F. Watts, 1974.

Machado de Assis, Joaquim Maria. *Epitaph of a Small Winner.* Trans. William Grossman. New York: Noonday, 1990.

Mainwaring, Scott. *The Catholic Church and Politics in Brazil, 1916-1985.* Stanford: Stanford University Press, 1986.

Martin, David. *Tongues of Fires: The Explosion of Protestantism in Latin America.* Oxford: B. Blackwell, 1990.

Mason, Tony. *Passion of the People?: Football in South America.* London: Verso, 1995.

Mattoso, Katia M. de Queirós. *To Be a Slave in Brazil, 1550-1888.* Trans. Arthur Goldhammer. New Brunswick, NJ: Rutgers University Press, 1986.

Maxwell, Kenneth. *Pombal: Paradox of the Enlightenment.* Cambridge: Cambridge University Press, 1995.

McGowan, Chris and Ricardo Pessanha. *The Brazilian Sound: Samba, Bossa Nova, and the Popular Music of Brazil.* New York: Billboard Books, 1991.

Morse, Richard. *The Bandeirantes: The Historical Role of the Brazilian Pathfinders.* New York: Alfred A. Knopf, 1965.

———. *New World Soundings: Culture and Ideology in the Americas.* Baltimore: Johns Hopkins University Press, 1989.

Nist, John A. *The Modernist Movement in Brazil: A Literary Study.* Austin: University of Texas Press, 1967.

Page, Joseph A. *The Brazilians.* Reading, MA: Addison-Wesley, 1995.

Pang, Eul-Soo. *Bahia in the First Brazilian Republic: Coronelismo and Oligarchies, 1889-1934.* Gainesville: University Presses of Florida, 1979.

Parker, Richard G. *Bodies, Pleasures, and Passions: Sexual Culture in Contemporary Brazil.* Boston: Beacon Press, 1991.

Pereira, Luiz Carlos Bresser. *Development and Crisis in Brazil, 1930-1983.* Trans. Marcia Van Dyke. Boulder, CO: Westview Press, 1984.

Perrone, Charles A. *Masters of Contemporary Brazilian Song: MPB, 1965-1985.* Austin: University of Texas Press, 1989.

Prado Júnior, Caio. *The Colonial Background of Modern Brazil.* Trans. Suzette Macedo. Berkeley: University of California Press, 1967.

Ramos, Graciliano. *Barren Lives.* Trans. Ralph Edward Dimmick. Austin: University of Texas Press, 1965.

Revkin, Andrew. *The Burning Season: The Murder of Chico Mendes and the Fight for the Amazon Rain Forest.* Boston: Houghton Mifflin, 1990.

Rodrigues, José Honório. *The Brazilians: Their Character and Aspirations.* Trans. Ralph Edward Dimmick. Austin: University of Texas Press, 1967.

Roett, Riordan. *Brazil: Politics in a Patrimonial Society,* 3rd ed. New York: Praeger, 1984.

Rosa, João Guimarães. *The Devil to Pay in the Backlands.* Trans. James L. Taylor and Harriet de Onís. New York: Alfred A. Knopf, 1963.

Schneider, Ronald. *"Order and Progress": A Political History of Brazil.* Boulder, CO: Westview Press, 1991.

Shoumatoff, Alex. *The Capital of Hope: Brasília and Its People.* New York: Vintage Books, 1980.

————. *The Rivers Amazon.* San Francisco: Sierra Club Books, 1978.

Skidmore, Thomas E. *Black Into White: Race and Nationality in Brazilian Thought.* New York: Oxford University Press, 1974.

————. *Politics in Brazil, 1930-1964: An Experiment in Democracy.* New York: Oxford University Press, 1967.

————. *The Politics of Military Rule in Brazil, 1964-85.* New York: Oxford University Press, 1988.

Souza, Márcio. *The Emperor of the Amazon.* Trans. Thomas Colchie. New York: Avon, 1977.

Stein, Stanley J. *Vassouras: A Brazilian Coffee County, 1850-1900.* Cambridge, MA: Harvard University Press, 1957.

Stein, Stanley J. and Barbara H. *The Colonial Heritage of Latin America: Essays on Economic Dependence in Perspective.* New York: Oxford University Press, 1970.

Stepan, Alfred. ed. *Democratizing Brazil: Problems of Transition and Consolidation.* New York: Oxford University Press, 1989.

Toplin, Robert Brent. *The Abolition of Slavery in Brazil.* New York: Atheneum, 1972.

Wagley, Charles. *An Introduction to Brazil,* rev. ed. New York: Columbia University Press, 1971.

Weschler, Lawrence. *A Miracle, A Universe: Settling Accounts with Torturers.* New York: Penguin Books, 1990.

Wesson, Robert G. *The United States and Brazil: Limits of Influence.* New York: Praeger, 1981.

Wirth, John D. *Minas Gerais in the Brazilian Federation, 1889-1937.* Stanford: Stanford University Press, 1977.

INDEX